Giorgio Guariso · Marialuisa Volta
Editors

Air Quality Integrated Assessment

A European Perspective

Editors
Giorgio Guariso
Politecnico di Milano
Milan
Italy

Marialuisa Volta
Università degli Studi di Brescia
Brescia
Italy

DOI 10.1007/978-3-319-33349-6

Library of Congress Control Number: 2016951968

Preface

The book reports in a handy but systematic way an extended survey across many European countries on the research activities and the current air quality plans at regional and local level. This allowed us to develop an Integrated Assessment Modelling (IAM) framework, to catalogue current approaches and to guide their implementation and evolution.

Integrated Assessment (IA) air pollution tools bring together data on pollutant sources (emission inventories), their contribution to atmospheric concentrations and human exposure, with information on emission reduction measures and their respective implementation costs. At the continental scale, such tools have been developed in the recent years to tackle these issues in a structured way. At the local/urban scale, however, only few IA systems have been developed and they have generally been used for non-reactive species. Thus, their application to suggest optimal policies to reduce secondary pollutants (i.e. those created in the atmosphere through chemical reactions of primary pollutants and currently those more affecting the air quality in European cities) has still relevant limitations.

The survey was performed within the APPRAISAL project (www.appraisal-fp7.eu) one of the projects of the 7th EU Framework Programme that analysed the situation and perspective of air pollution management in Europe. In particular, APPRAISAL's survey was aimed at understanding the degree at which the Integrated Assessment approach to air quality problems is adopted by regional authorities, on the one side, and researchers, on the other. More precisely, it involved the following:

- a review of the modelling methodologies in place across EU member states to identify sources and to assess the effectiveness of emission reduction measures at all scales (including downscaling of impacts to city level which are a main concern with respect to compliance with the requested limit values),
- a review of the methodologies to assess the effects of local and regional emission abatement measures on human health,
- a review of monitoring data and complementary methodologies, e.g. source apportionment, to identify their potential synergies in a general integrated assessment frame,

- a review of the techniques used to evaluate the robustness and uncertainties of the assessment,
- an analysis of the emission abatement policies and measures planned at regional and local scales,
- their synergies/trade-offs with the measures implemented at the national scales (e.g. national emissions ceilings or national climate change programmes).

These tasks have been performed by defining a common and structured format, i.e. by designing a database and populate it, clearly specifying the meaning of all keywords in order to guarantee a uniform understanding across all countries and applications. A collaborative multiple user tool has been implemented to allow all involved agencies to fill the questionnaire through a Web application.

The project has been the result of the cooperation of 16 research groups in nine different European countries (see Figure) with the contribution of six stakeholders (local environmental authorities of different EU regions), and with the collaboration of FAIRMODE (the Forum for air quality modelling in Europe, http://fairmode.jrc.ec.europa.eu/) and NIAM (Network for Integrated Assessment Modelling, http://www.niam.scarp.se/) initiatives.

The project lasted from 2011 to 2014 and was coordinated by the University of Brescia, Italy.

The material produced by all the project activities is available online on the project website. The content of this book is largely drawn from the project deliverables.

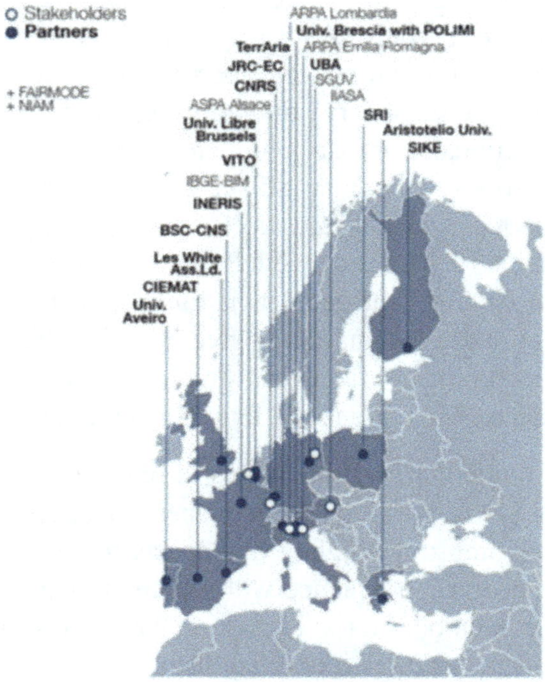

Milan, Italy G. Guariso
Brescia, Italy M. Volta

Contents

Chapter 1
Air Quality in Europe: Today and Tomorrow

G. Guariso and M. Volta

The last "Air quality in Europe" report by the European Environmental Agency (EEA 2015) foresees almost five millions of years of life lost (YOLL) in the 28 EU Member States due to the high concentrations of PM2.5. YOLLs are an estimate of the average years that a person would have lived if he or she had not died prematurely, giving greater weight to deaths at a younger age and lower weight to deaths at an older age. For the 507.4 million inhabitants of EU, this means an average loss of more than 3 days each year.

Furthermore, speaking about the average conditions, for air quality has a limited meaning. The situation is normally worse in highly populated areas where most population lives and, for the same reason, emission of pollutant are higher.

Indeed, the same report, referring to 990 urban monitoring stations in 736 European cities, shows that 202 of them (27.4 %) have exceeded the limit of 35 days above 50 μg/m^3 for PM10 average daily concentrations.

The situation is quite different in different EU Member States (MS) and within each MS. Figure 1.1 shows for instance the 36-th highest daily mean and the 25 and 75 % percentiles (box limits) in each MS compared to the European limit of 50 μg/m^3. As we will see in the following chapters, exact links between pollutant concentrations and health impacts are not completely known and thus the limits proposed by the World Health Organization are even stricter than those adopted by EU regulations.

Figure 1.2 expresses this situation in geographical terms, showing where the exceedance of the EU limit for PM10 is reported.

The situation is quite similar for other traditional pollutant such as NOx and only slightly more complex for Ozone, as shown in Fig. 1.3.

G. Guariso
Politecnico di Milano, Milan, Italy

M. Volta (✉)
Università degli Studi di Brescia, Brescia, Italy
e-mail: marialuisa.volta@unibs.it

G. Guariso and M. Volta (eds.), *Air Quality Integrated Assessment*,
PoliMI SpringerBriefs, DOI 10.1007/978-3-319-33349-6_1

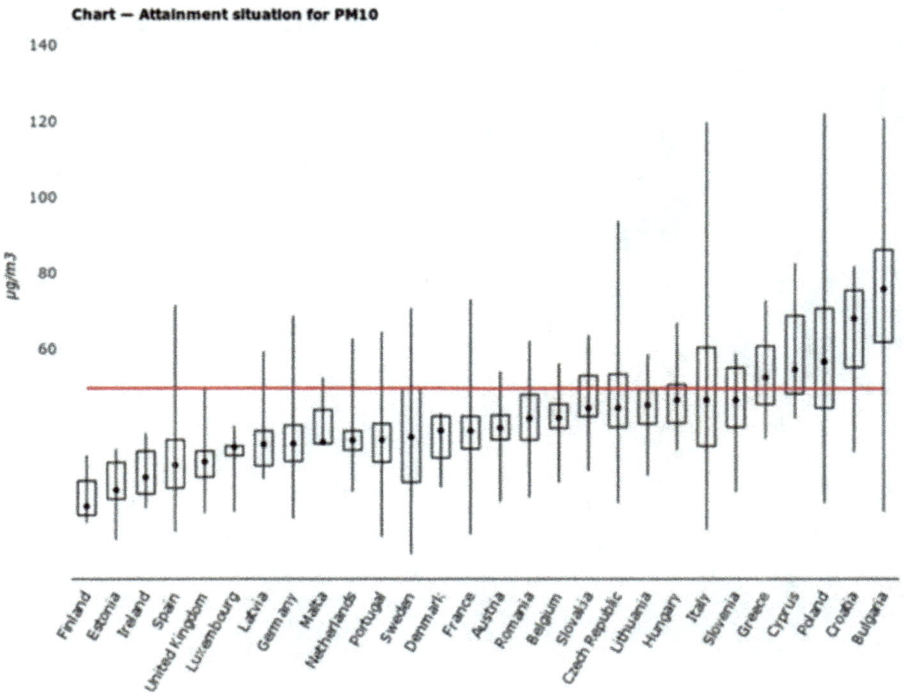

Fig. 1.1 Distributions of the 36-th highest PM10 daily value in EU MS (*source* EEA 2015)

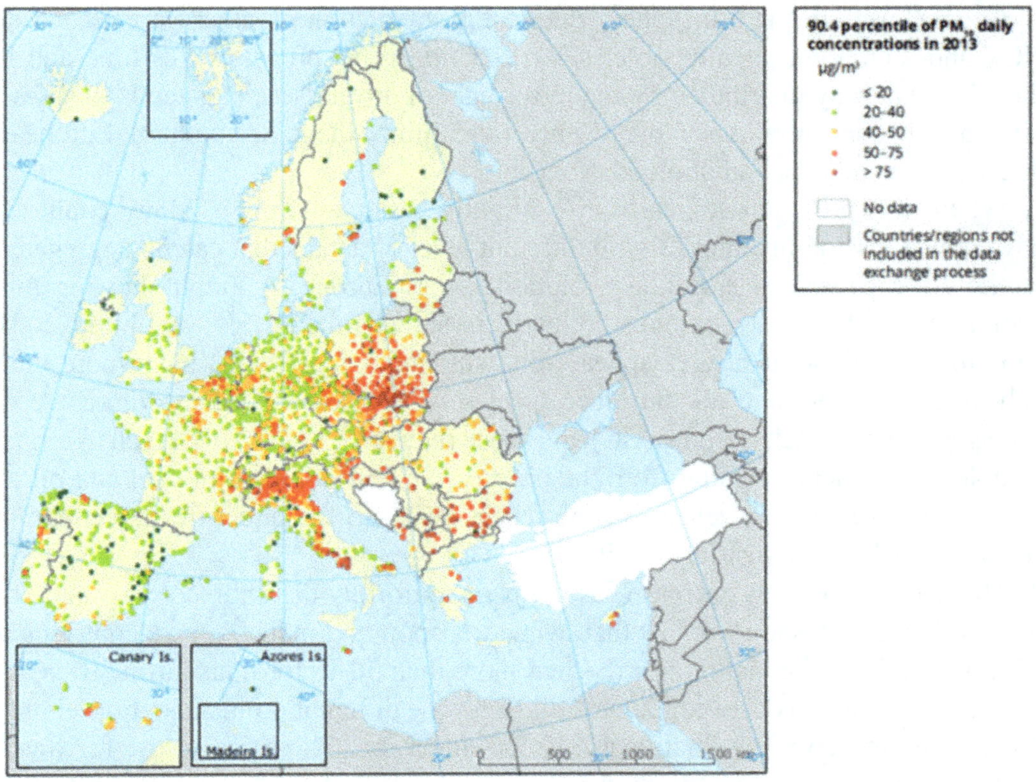

Fig. 1.2 Geographical distribution of the 36-th highest PM10 daily value (*source* EEA 2015)

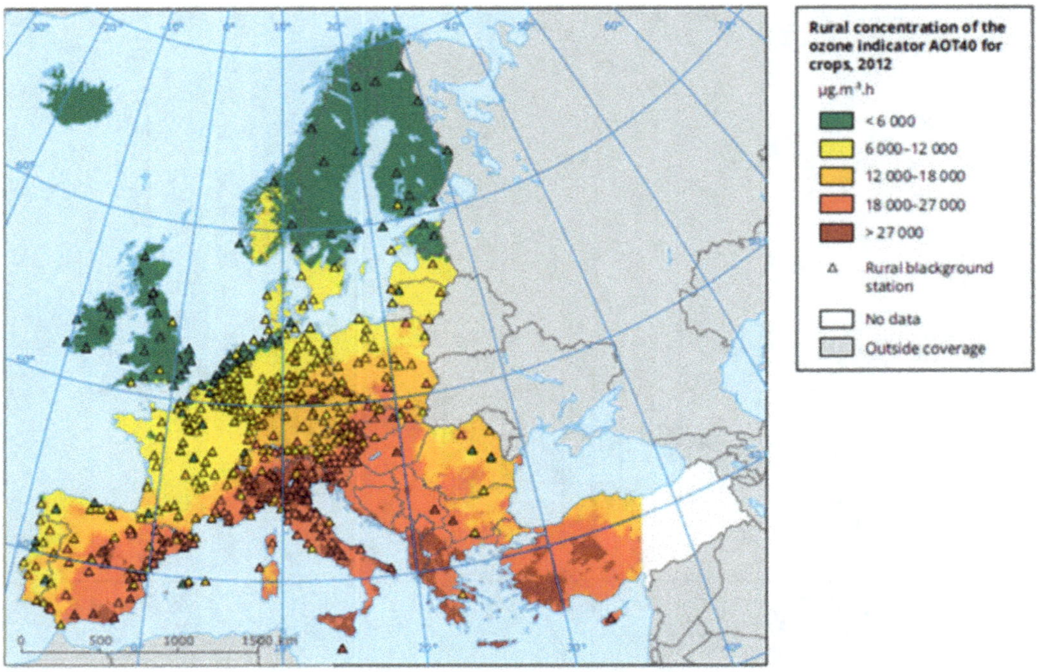

Fig. 1.3 Geographical distribution of AOT40, an indicator of air quality impacts on crops (*source* EEA 2015)

Ozone forms in the atmosphere due to the interaction of other gases (such as NOx and VOC) and of ultraviolet solar radiation. This process takes time and is therefore naturally distributed by the movement of air masses. This tends to spread high ozone concentrations more evenly (and limits them to southern European countries where solar radiation is stronger).

Where this pollution comes from is slightly easier to explain. Many countries have now emission inventories with different level of details that can be aggregated to show the pattern of emission evolution across Europe. A graph showing this evolution for the most common pollutant is shown in Fig. 1.4, assuming 2004 emission as 100 %. It clearly appears that sulphur oxides (SOx) have more than halved in ten years and all the other species have also reduced in different percentages, being black carbon (BC) the least reduced (5 %). This results from a complex set of actions going from the progressive abandonment of coal and oil as fuels to turn to gas, as well as, in the recent years, to the effect of the economic crisis that reduced industrial activities.

The above emissions decrease has not been uniformly distributed across activity sectors. Figure 1.5 shows in fact that, while transport and industry have contributed a lot (the emission reduction has reached more than 50 % for transport in 10 years and that of industry is between 20 and 40 % for the different pollutants), households and agriculture have been stationary, if not increasing. The same is true for waste treatment, even if the contribution of this sector to the total emission budget is small, except for CH_4. Finally, the contribution of the energy sector is somehow mixed: most pollutants have decreased (NOx, for instance, by more than 70 %)

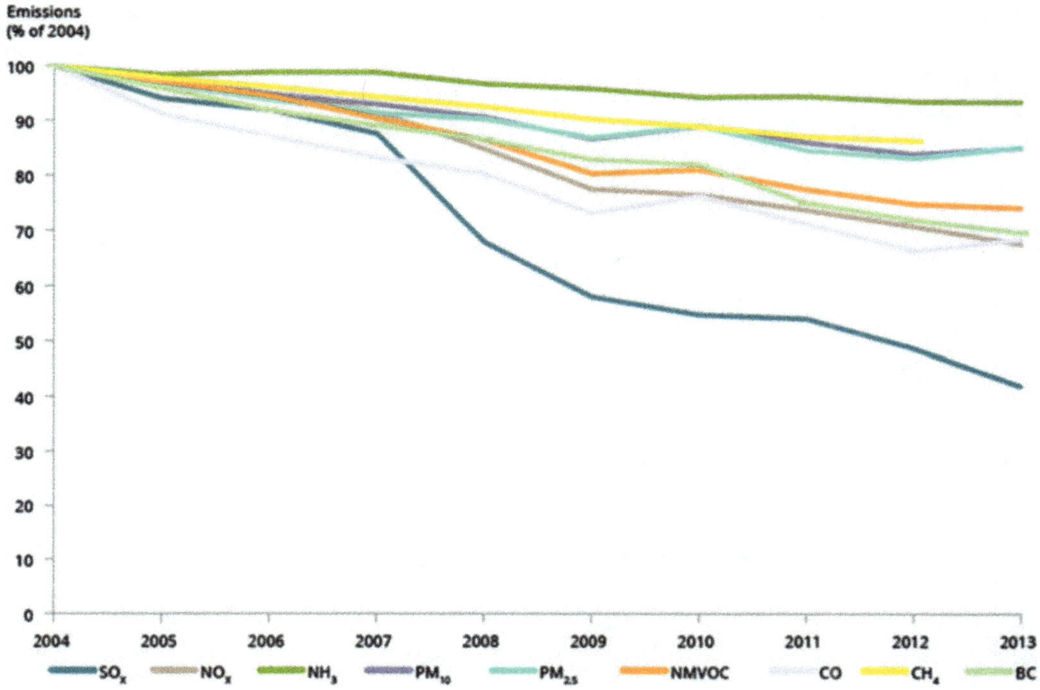

Fig. 1.4 Evolution of EU pollutant emissions through time (2004 = 100 %) (*source* EEA 2015)

while others, like primary PM, have slightly increased, possibly because of the increased use of biomass burning.

When talking of a large territory (Europe, a country, a region within a country) the link between the perceived pollution (the concentration, that causes adverse effects) and its causes (the emission) is not straightforward. Two aspects must in fact be considered and play an essential role in defining such a link: the meteorology and the chemistry of the atmosphere. Meteorology obviously determines if a certain emission remains more or less confined in the air above the emission source or is dispersed far away from it. In the first case, the concentration may reach very high values, in the second the source contribution may become negligible. Whether in the first or in the second case (and in all intermediate situations), it depends on the climate and orography of each specific area. Along the seashores or at the foot of the mountains, there are always breezes that may move the air masses, while there are flat areas where wind speed is always extremely low.

The second aspect is the chemistry of the atmosphere. Most pollutants are indeed reactive and, when entering the atmosphere, they start combining with other components and producing different substances. While for some pollutants, say for instance SO_2, such processes can be so slow to be negligible in most cases, for other substances, like NOx or VOC, they take place in time of hours and thus must be accurately considered. For instance, a component more or less relevant of PM (it depends on the local chemistry of the atmosphere) and tropospheric ozone are secondary pollutants, meaning that they are not directly emitted, but formed in the atmosphere due to the specific conditions and the presence of other gases, called

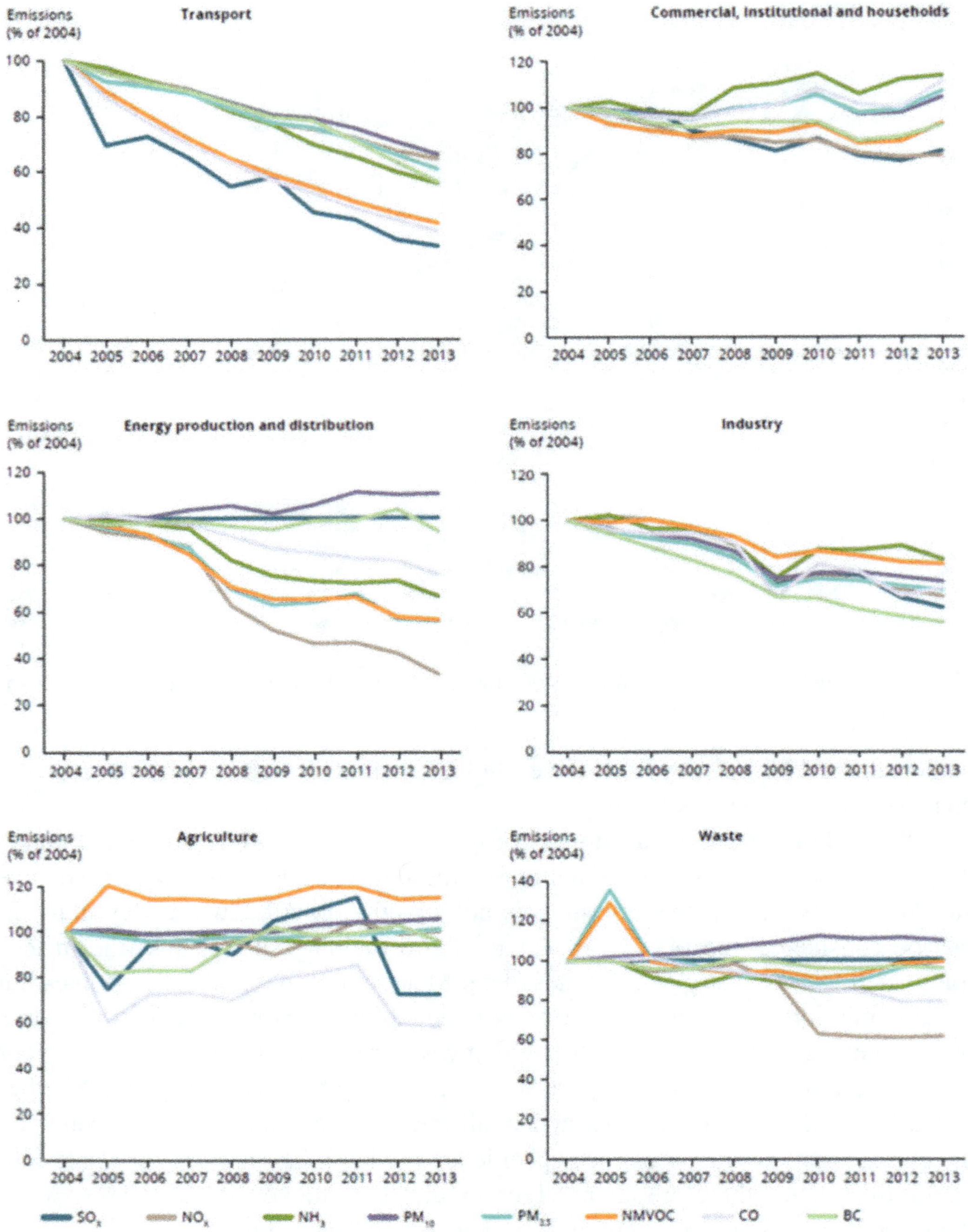

Fig. 1.5 Evolution of pollutant emissions in different sectors (2004 = 100 %) (*source* EEA 2015)

"precursors". Since they represent by far the most dangerous pollutant in EU today, working for their reduction is extremely complex since the problem must be tackled considering a large area and not a single source and that one has to operate on the precursors, knowing that meteorology may alter the picture in different ways.

Given this complex situation, EU has issued a number of directives to define limits concentration on ambient air and indications on how to attain such results. As is apparent from the preamble to Air Quality Directive 2008/50/EC (AQD),

European air quality legislation puts the main emphasis on protecting human health and the environment as a whole and stresses that "it is particularly important to combat emissions of pollutants at source and to identify and implement the most effective emission reduction measures at local, national and Community level." These basic principles have already been formulated in the former so-called air quality framework directive (96/62/EC) and its daughter directives (1999/30/EC, 2000/69/EC, 2002/3/EC, 2004/1 007/EC).

The set of actions foreseen by the current legislation (CLE) is expected to continue the reduction of emissions of the past decade and thus to bring a general improvement for the decade to come. Despite this, some urban areas and some regions will still struggle with severe air quality problems and related health effects. These areas are often characterized by specific environmental and anthropogenic factors and will require ad hoc additional local actions to complement medium and long term national and EU-wide strategies to reach EU air quality objectives. At the same time, these urban areas are among the territories where most energy is consumed and most greenhouse gases (GHGs) are emitted. The reviews of the Thematic Strategy on Air Pollution (Amann et al. 2011; Kiesewetter et al. 2013) have used the European air pollution model GAINS to study the trends of compliance evolution from the base year 2010–2025 (assuming current legislation only), the improvement for a 2025 scenario and the further compliance achieved in 2030 by implementing all technical measures (Maximum Technically Feasible emission Reductions, MTFR). The assessment of compliance of the daily PM10 exceedances limit value with respect to the current Ambient Air Quality Directive is shown in Fig. 1.6.

Some important observations can be derived from these figures.

Comparing the 2010 map with the 2025 CLE case, it clearly appears the move away from a general picture of non-compliance (2010) to few limited remaining areas of non-compliance. European wide measures (already mandated) will determine a significant improvement in compliance especially in the old EU-15 Member States. What is also clear by comparing the 2025 CLE with the 2025 A5 (defined as 'central policy scenario') is the limited potential of further EU-wide measures to improve compliance; this is further underlined by the 2030 MTFR scenario, that shows still various areas of uncertain or unlikely compliance even when adopting all the available abatement technologies.

Introducing tougher European-wide measures to address residual non-compliance confined to 10 % of the urban zones in Europe would likely be significantly more costly than directly addressing these areas with specifically designed measures based on bottom-up Integrated Assessment (IA) approach using regional/local data. In this regard, regional IA software tools such as RIAT (Carnevale et al. 2012), LEAQ (Zachary et al. 2011), etc. with their ability to identify cost-optimised local strategies are already available to quantify the cost-effective split between further European wide measures and regional/local measures. They will inevitably find wider application and play an increasing role in these emerging 'discrete islands of non-compliance'.

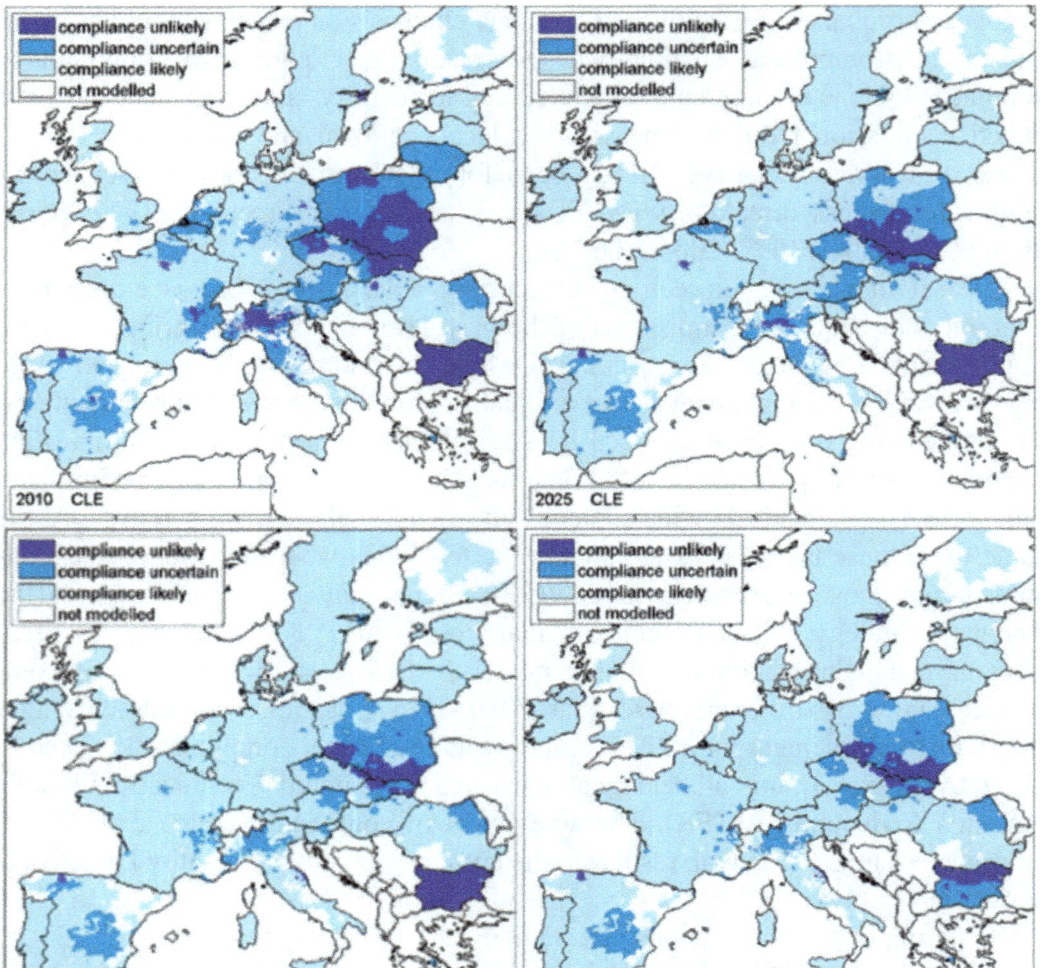

Fig. 1.6 Evolution of PM10 compliance according to GAINS results (*source* Amann 2013)

These observations motivate the growing interest in IA models and tools for local and regional scale. Their importance became apparent again in connection with Article 22 of AQD 2008 "Postponement of attainment deadlines and exemption from the obligation to apply certain limit values" commonly called "notification for time extension". For both air quality plans and time extension, more elaborated requirements are formulated in Annex XV compared to former regulations. The implementing decision of December 2011 (201 1/850/EU) reflects this, clearly looking at the reporting obligations laid down there (Article 13, Annex II, Section H, I, J and especially K) and looking at the amount of information that has to be provided regularly (e-reporting has entered full operation mode from January 2014). Finally, "Air quality plans" according to AQD Art. 23 are the strategic element to be developed, with the aim to reliably meet ambient air quality standards in a cost-effective way.

This growing set of developments and activities required to be framed and organized to allow better understanding of the different approaches in use, to be able to compare their characteristics, and ultimately to suggest how to diffuse best practices and in which direction to move additional research and new software implementations. The ultimate scope of such effort that is briefly summarized in the following chapters is to provide decision-makers in charge of air pollution management with a view of the current European situation and a way of improving their current policies.

Acknowledgments This chapter is partly taken from APPRAISAL Deliverable D2.2 (downloadable from the project website http://www.appraisal-fp7.eu/site/documentation/deliverables.html).

References

Amann M (ed) (2013) Policy scenarios for the revision of the thematic strategy on air pollution, TSAP Report #10, Version 1.2, IIASA, Laxenburg

Amann M (ed) (2014) The final policy scenarios of the EU Clean Air Policy Package, TSAP Report #11. IIASA, Laxenburg

Amann M, Bertok I, Borken-Kleefeld J, Cofala J, Heyes C, Höglund-Isaksson L, Klimont Z, Nguyen B, Posch M, Rafaj P, Sandler R, Schöpp W, Wagner F, Winiwarter W (2011) Cost-effective control of air quality and greenhouse gases in Europe: modelling and policy applications. Environ Model Softw 26:1489–1501

Carnevale C, Finzi G, Pisoni E, Volta M, Guariso G, Gianfreda R, Maffeis G, Thunis P, White L, Triacchini G (2012) An integrated assessment tool to define effective air quality policies at regional scale. Environ Model Softw 38:306–315

EEA (2015) Air quality in Europe—2015 report. Publications Office of the European Union, Luxembourg

Kiesewetter G, Borken-Kleefeld J, Schöpp W, Heyes C, Bertok I, Thunis P, Bessagnet B, Terrenoire E, Amann M (2013) Modelling compliance with NO_2 and PM10 air quality limit values in the GAINS model. TSAP Report #9, IIASA, Laxenburg

Zachary DS, Drouet L, Leopold U, Aleluia Reis L (2011) Trade-offs between energy cost and health impact in a regional coupled energy–air quality model: the LEAQ model. Environ Res Lett 6:1–9

Chapter 2
A Framework for Integrated Assessment Modelling

N. Blond, C. Carnevale, J. Douros, G. Finzi, G. Guariso, S. Janssen, G. Maffeis, A. Martilli, E. Pisoni, E. Real, E. Turrini, P. Viaene and M. Volta

2.1 Introduction

"Air quality plans" according to Air Quality Directive 2008/50/EC Art. 23 are the strategic element to be developed, with the aim to reliably meet ambient air quality standards in a cost-effective way. This chapter provides a general framework to develop and assess such plans along the lines of the European Commission's basic

N. Blond
Centre National de la Recherche Scientifique (CNRS), Paris, France

C. Carnevale · G. Finzi · E. Turrini · M. Volta (✉)
Università degli Studi di Brescia, Brescia, Italy
e-mail: marialuisa.volta@unibs.it

J. Douros
Aristotle University of Thessaloniki, Thessaloniki, Greece

G. Guariso
Politecnico di Milano, Milan, Italy

S. Janssen · P. Viaene
Vlaamse Instelling Voor Technologisch Onderzoek N.V. (VITO), Mol, Belgium

G. Maffeis
TerrAria srl, Milan, Italy

A. Martilli
Centro de Investigaciones Energeticas, Medioambientales y Tecnologicas (CIEMAT), Madrid, Spain

E. Pisoni
European Commission, Joint Research Centre (JRC), Directorate for Energy, Transport and Climate, Air and Climate Unit, Ispra, Italy

E. Real
Institut National de l'Environnement et des Risques (INERIS), Verneuil-en-Halatte, France

G. Guariso and M. Volta (eds.), *Air Quality Integrated Assessment*,
PoliMI SpringerBriefs, DOI 10.1007/978-3-319-33349-6_2

ideas to implement effective emission reduction measures at local, regional, and national level. This methodological point of view also allows to analyse the existing integrated approaches.

2.1.1 The DPSIR Framework Concept

To comply with the above aims requires the key elements of an Integrated Assessment Modelling (IAM) approach to be carefully defined. These elements will be derived by the general EEA DPSIR scheme (EEA 2012) and a holistic approach. The overall framework should:

- Be structured in a modular way, with data flows connecting each building block;
- Be interconnected to higher decision levels (i.e. national and European scales);
- Consider the approaches available to evaluate IAM variability (taking into account both the concept of "uncertainty", that is related to "variables/model results" that can be compared with real data, and the concept of "indefiniteness", related to the impacts of future policy decisions)
- Be sufficiently general to include the current experiences/approaches (presented in the next chapter) and,
- Show, for each module of the framework, different "levels of implementation complexity".

The last two points are quite important. The idea is that, looking at the different "levels of complexity" defined for each DPSIR block, one should be able to grasp in which "direction" to move to improve the detail (and, hopefully, the quality) of his own IAM implementation. This should translate into the possibility to assess the pros and cons for enhancing the level of detail of the description of each block in a given IAM implementation, and thus compare possible improvement with the related effort. The final idea is to be able to classify existing European plans and projects, with the aim not to provide an assessment value of the plans themselves, but to show possible "directions" of improvement, for each building block of each plan.

In the next section, at first, a general overview of the proposed framework will be provided. Then, each building block will be described in detail, focusing on input, functionality, output, synergies among scales, and uncertainty and defining three possible tiers of different complexity.

2.2 A General Overview of the IAM Framework

The DPSIR analytical concept (Fig. 2.1) is the causal framework for describing the interactions between society and environment, adopted by the European Environment Agency. The building blocks of this scheme are:

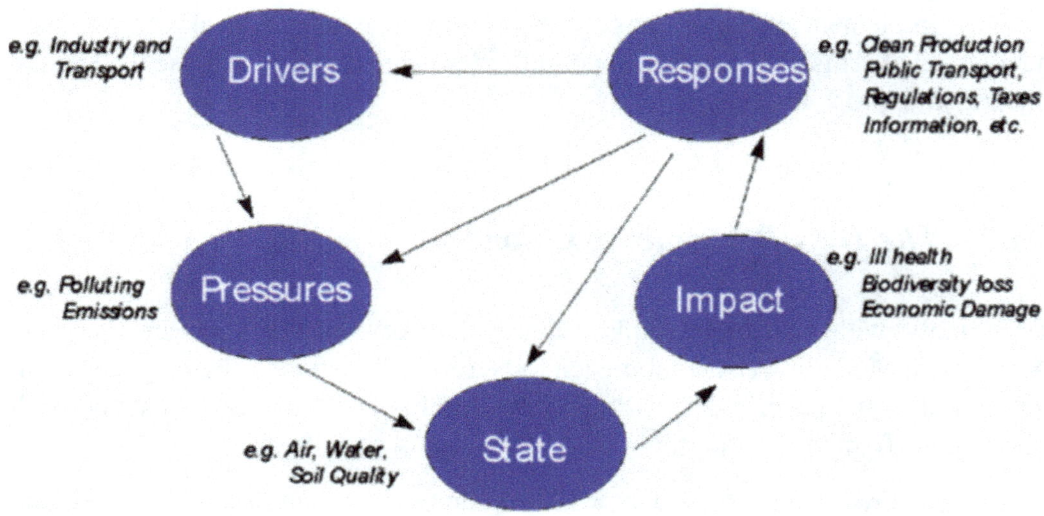

Fig. 2.1 The general DPSIR scheme (*source* http://www.eea.europa.eu/)

- DRIVING FORCES,
- PRESSURES,
- STATE,
- IMPACT,
- RESPONSES,

and represent an extension of the PSR model developed by OECD (definitions from *EEA glossary*, available at http://glossary.eea.europa.eu).

The DPSIR scheme helps "to structure thinking about the interplay between the environment and socioeconomic activities", and "support in designing assessments, identifying indicators, and communicating results" (EEA 2012). Furthermore, a set of DPSIR indicators has been proposed, that helps to reduce efforts for collecting data and information by focusing on a few elements, and to make data comparable between institutions and countries. Starting from these definitions and features, it has been decided to adapt the DPSIR scheme to IAM at regional/local scale (considering with this definition domains of few hundreds kilometres). So the DPSIR scheme shown in Fig. 2.1 has been translated into the framework illustrated in Fig. 2.2.

In particular, in the scheme in Fig. 2.2, the meaning of each block is as follows (quoting again from *EEA glossary*):

- **DRIVERS**: this block describes the "actions resulting from or influenced by human/natural activity or intervention". Here we refer to variables (often called "activity levels") describing traffic, industries, residential heating, etc.
- **PRESSURES** (Emissions): this block describes the "discharge of pollutants into the atmosphere from stationary sources such as smokestacks, and from surface areas of commercial or industrial facilities and mobile sources, for example, motor vehicles, locomotives and aircrafts." PRESSURES depend on DRIVERS, and are computed as function of the activity levels and the quantity of pollution emitted per activity unit (emission factor).

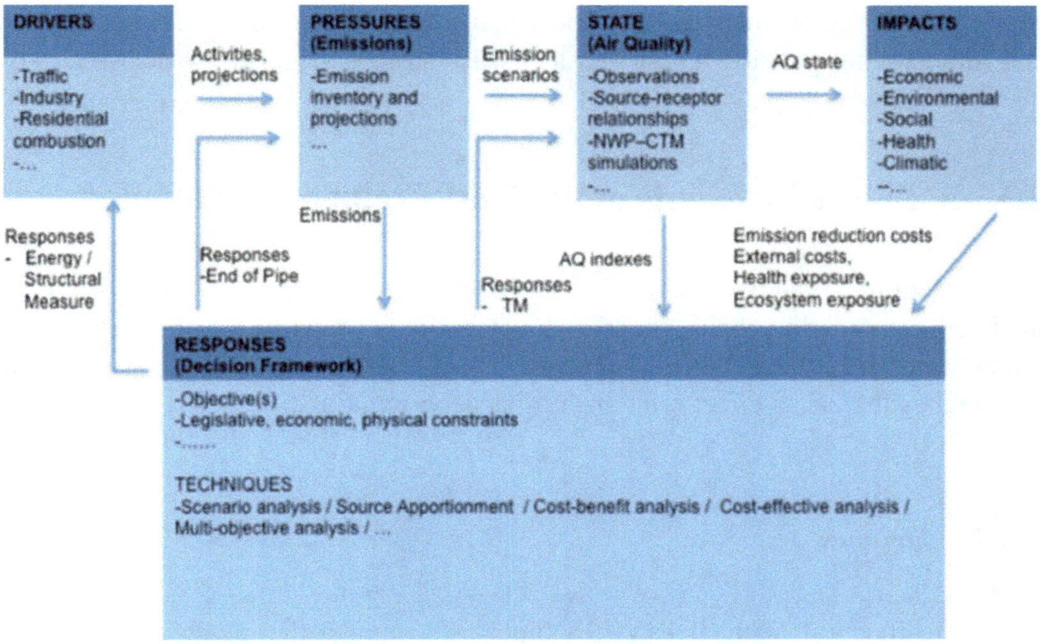

Fig. 2.2 The DPSIR scheme adapted to IAM of air quality at regional/local scale

- **STATE** (Air quality): this block describes the "condition of different environmental compartments and systems". Here, we refer to STATE as the concentrations of air pollutants resulting from the PRESSURES defined in the previous block. In IAM implementations, STATE can sometimes be directly measured, but more often it is computed using some kind of air quality model.
- **IMPACT**: this block describes "any alteration of environmental conditions or creation of a new set of environmental conditions, adverse or beneficial, caused or induced by the action or set of actions under consideration". In the proposed framework, we refer to IMPACT on human health, vegetation, ecosystem, etc. derived by a modification of the STATE. Again the calculation of the IMPACT may be based on some measure, but normally requires a set of models (e.g. health impacts are often evaluated using dose-response functions).
- **RESPONSES**: this block describes the "attempts to prevent, compensate, ameliorate or adapt to changes in the state of the environment". In our framework, this block describes all the measures that could be applied, at a regional/local scale, to improve the STATE and reduce IMPACT.

It is worthwhile to note that the scheme in Fig. 2.2 is integrated with "higher" decision levels. This means that for each block some information is provided by "external" (not described in the scheme) components. For instance, the variables under DRIVERS may depend on GDP growth, population dynamics, etc.; the STATE may also depend on pollution coming from other regions/states; or the RESPONSES may be constrained by economic factors. Each block can thus be seen as receiving external forcing inputs that are not shown explicitly in Fig. 2.2, since they cannot be influenced (or just marginally) by the actions under consideration.

More specifically, all regional and local plans are to be compatible with national and international policies. These "scale" issues are discussed in the next sections.

2.3 A Detailed Analysis of the IAM Framework Modules

In this section, all the five building blocks of the IAM framework will be discussed in detail, considering their "input", "functionality", "output", "synergies among scales" and "uncertainty". The "functionality" is the core part of the description, and defines the cause-effect relationship between input and output.

2.3.1 Drivers

The basic function of the DRIVERS block is to model the development of key driving activities (i.e. road traffic, off-road traffic and machinery, residential combustion, centralized energy production, industry, agriculture) over time (Amann et al. 2011). It thereby provides input to the PRESSURES block in the form of, e.g., road traffic kilometres driven, residential heating fuel consumption, etc. (dis)aggregated in such a way that it includes emission-wise relevant classification of sectors, sources and technologies.

To provide relevant information to the PRESSURES block, DRIVERS have to be quantified with specific measurable variables. For instance, special attention has been given in European plans to the sectors that are important for urban air quality (road traffic, residential heating, industry). The next Table 2.1 gives an overview of the most important activity parameters used to quantify each of these sectors.

Input

Input parameters are factors that represent causes of emission-wise essential activities. Important input parameters include general factors such as population, general economic activities (e.g. in the form of GDP), more specific activity factors (e.g. sector specific production intensities, transport demand, energy demand etc.) and technology change factors (e.g. vehicle stock structure, energy efficiency of buildings etc.) that may be driven by international, national or local requirements or "natural", non-forced development.

Table 2.1 Parameters commonly used to quantify relevant urban activities

Sector	Key activity parameters
Road traffic	Kilometres driven, fuel consumption
Off-road and machinery	Fuel consumption
Residential combustion	Fuel consumption, heat production
Energy production and industry	Fuel consumption, energy/industrial production

Functionality
The functionality expresses the cause-effect relationship (or model) between the input and the output, e.g. considering how transport demand of goods and people translates into kilometres driven and/or fuels used in different types of vehicles. While for some "base" period (often a past year for which a fairly complete set of data exists) an inventory is often adequate to attain directly the output of the DRIVERS block (e.g. transport kilometres driven or fuel used), for projections into the future the input-functionality-output chain needs to respond to the assumed future changes in economic activities, technology developments, etc. This chain can be implemented at different levels of complexity, from simple calculation of cause-effect relationships to detailed traffic, housing and energy system models. City or regional level assessments can be implemented using local information (bottom-up), or derived from national level models (top-down), or as a combination of both approaches. Models with dynamic spatial capabilities are desirable to be able to assess changes in spatial patterns of activities.

In general, for the DRIVERS block implementation, the following three-level classification can be adopted:

- **LEVEL 1**: when a top-down approach is applied, using coarse spatial and temporal allocation schemes;
- **LEVEL 2**: when a bottom-up approach with generic (i.e. national/aggregated) assumptions is applied, using realistic spatial and temporal allocation schemes;
- **LEVEL 3**: when a bottom-up approach with specific (i.e. local/detailed) assumptions is applied, using local spatial and temporal allocation schemes.

In the following sections, a more detailed description of the DRIVERS block implementation will be provided, focusing on two important aspects of DRIVERS, that is to say:

- Base year inventory and projections;
- Spatial and temporal assessment.

Base Year Inventory and Projections
The inventory of activities and emission-wise relevant technologies can be based on the data collected or modelled from the respective city area or region (bottom-up approach), or on statistics of a wider area (typically a country) of which the share of the respective city area or region is defined using weighting surrogates (top-down approach).

In some cases it might be difficult to attain reliable, representative collected data from certain areas. For instance, technology stock inventory at sub-national level is often not practical, and national level data have to be used. In case of a top-down approach, the reliability of the activity estimate depends on the representativeness of the weighting surrogates used.

For future projections, it is particularly important that the changes in time of the input of the DRIVERS block (e.g. changes in population, economical activities, transport needs etc.) realistically translate into output (i.e. activities and technologies). Therefore the assessment of future developments of the DRIVERS block typically requires a more sophisticated framework than what would be needed for the base year inventory.

In the following, the main emission source sectors are discussed in addition to the general three-level approach presented above.

Road traffic activities and projections are typically relatively well known at city level because these data are of interest also for other bodies than environmental assessment. In addition to factors affecting tail-pipe emissions, non-exhaust road dust emissions are an important impairer of air quality. Important parameters for non-exhaust emission factors, in addition to vehicle types, are tire type, road surface type and climate conditions. Transport demand based modelling approaches enable also assessment of spatial changes.

The three tiers classification presented above may be represented, for instance, by:

1. Allocation of traffic activity data from national level (top-down). The allocation may be based on population data (in relation to national total);
2. Activities based on city level traffic counts or other estimate (bottom-up), and allocation of vehicle categories and technologies based on national average (top-down);
3. Activities based on city level traffic counts or other estimate, distinguished for each vehicle category and technology using city level survey data (bottom-up) or other local data (e.g. city level traffic model).

Availability of activity data for **off-road traffic and machinery** is variable. For sea vessels, trains and airplanes, activities often are relatively well known. On the other hand, activity data can be much more uncertain for construction and maintenance machinery activities derived from national level because of the lack of appropriate weighting surrogates. However, reliable estimate on the changes in vehicle stock age structure is essential especially for traffic and machinery because of remarkable differences in emissions factors of various EURO standard levels. The level of complexity might be similar to that of road traffic taking into account that for each specific category (rail traffic, aviation, marine, harbours, military, agriculture machinery, industry, construction, maintenance, etc.…) different proxy variables must be used.

Residential combustion activities are often relatively uncertain. Especially for residential wood combustion, which is a major concern from air quality perspective in many European cities because of its high fine particle emissions, bottom-up approaches can rarely be based on sale statistics because a lot of wood fuel is used privately. For future changes, several factors should be taken into account: competitiveness of different heating systems, prospects of citizens' preferences, renewal of heating appliance stock and its effect on emission factors, changes in fuel qualities, legal requirements (e.g. Eco-Design Directive). The use of detailed housing and/or zoning models could enable the assessment of spatial changes in the future.

In case there is no reliable estimate of local level activity or practicable procedure for top-down allocation, source apportionment techniques might be considered to detect an initial order-of-magnitude evaluation of the residential combustion activities.

Once again, the three-level classification may be characterized by:

1. Allocation from national level values (top-down). The allocation may be based on surrogate data representing residential combustion activity in a coarse manner, e.g., number of residential houses or population data (in relation to national total);
2. Based on city level estimates about respective activity (e.g. local sales statistics of fuels or surveys about fuel use), or allocated from national data using surrogates that represent residential combustion activity more realistically (e.g. average fuel use per household for different types of houses). Projections can be based on city level residential combustion for each fuel/heating type (bottom-up);
3. Activities distinguished for each house type and/or combustion technology categories using city level survey data (bottom-up) or other locally specific data (e.g. city level building heating/cooling model).

For **large energy production and industrial plants**, activity and technology information can be sometimes attained even at individual plant or process level. For projections, factors such as new plant or technology investments, agreed plants shut-offs, local level goals and agreements on e.g. renewable energy, effects of national level prospects in energy production and industry, changes in legal requirements (e.g. IE Directive) etc. should be taken into account.

The three-level classification may be given by:

1. Allocation from national level of energy/industrial production activity for each fuel/industrial product (top-down). The allocation may be based on production capacity or annual production (in relation to national total) and information about national averages of production and emission control technologies;
2. Based on local level total energy/industrial production activity amounts for each fuel/industrial product and information about production and emission control technologies data at local level;
3. Based on individual plant data about energy/industrial production activity amounts as well as production and emission control technologies.

Agriculture emissions are often disregarded in urban assessments. However, at national level, agriculture is often the major source of ammonia emissions and can be relatively important in PM emissions. Base year data include animal numbers, use of different types of animal houses and their ventilation and air treatment technologies, different manure application methods etc. Projections typically include development of animal numbers following national agriculture policies and/or market prospects of agricultural products.

Spatial and Temporal Assessment

To provide appropriate information to the PRESSURE block, it is important to know not only the quantity but also the physical location and temporal variation of emission releases. Therefore, in order to be able to resolve the emissions in space and time, the activities (i.e. the DRIVERS block) must be allocated to certain grid and temporal patterns. The spatial aspect is particularly important in city or local level assessments for local emissions may cause considerable impacts on human populations.

The spatial allocation of point sources simply implies the association of the geographical location and height of the stack with the corresponding grid cell and vertical layer of the atmospheric model, respectively. Area emissions, by contrast, must be spatially allocated using again weighting factors, i.e. surrogates. The choice of surrogate parameters for different source sectors depends on the availability of data that would represent the emission distribution in a given sector at the desired spatial resolution. The temporal variation for different sectors can be based on internationally, nationally or locally defined default variations or local data (e.g. questionnaires or observed data). The following provides a proposal for three levels of complexity in spatial and temporal assessment for different source sectors.

Road traffic network is typically available for spatial allocation. To distinguish between more or less busy roads and different driving conditions, availability of data may vary. Non-exhaust emissions vary highly in space and time depending also on other factors than driving amounts and conditions or vehicle technology (e.g. road surface type and condition, seasonal and hourly climate conditions). These factors might be difficult to take into account with a reasonable accuracy without specific road dust models.

A three-level classification might be:

1. Spatial assessment based on road network data with coarse traffic allocation scheme (e.g. using road type classification to distinguish more and less trafficked roads). Temporal variation based on general default variations.
2. Spatial assessment based on road network data with more realistic representation of traffic flows (e.g. actual traffic counts for each road segment). Temporal variation based on nationally or locally defined default variations.
3. Spatial assessment based on road network data with representation of district traffic flows for vehicle categories and/or driving conditions (e.g. based on a city level traffic model). Traffic demand based modelling approaches are desirable to assess spatial changes in future projections. Temporal variation should be based on locally observed data.

Data availability for spatial allocation of **off-road traffic and machinery** is variable. For some forms, the locations of activities are relatively well known, e.g. for sea vessels, trains and airplanes. For many forms of machinery, in contrast, the basis for spatial allocation can be much more complex.

Three-level classification:

1. Coarse spatial allocation scheme for each off-road and machinery sub-categories (e.g. gridding based on land use data about aviation, harbour, military, agricultural, industrial areas, population data, etc.). Temporal variation based on general national default variations.
2. Spatial allocation with more realistic representation of activity for each off-road and machinery sub-categories (e.g. gridding with estimate about the location of activity inside respective land-use classes). Temporal variation based on nationally or locally defined default variations.
3. Spatial allocation for each off-road and machinery sub-categories based on activity intensities in respective locations (e.g. based on train/aircraft/vessel movements, GPS data and/or activity model). Temporal variation based on locally observed data.

Residential combustion activities are often poorly registered, because in many countries/cities individual household level heating systems do not need licenses. Therefore spatial allocation has to be based on some more general household level data, e.g. building registers.

Three-level classification:

1. Coarse spatial allocation scheme for each residential heating fuels and/or main heating sub-categories (e.g. gridding based GIS data on number of residential houses or population data). Temporal variation based on general default variations.
2. Spatial allocation with more realistic representation of activity for each residential heating fuels and/or main heating sub-categories (e.g. gridding based on GIS data on number or floor area of different types of buildings or other relevant information that distinguishes residential fuel use intensities in different building types). Temporal variation based on nationally or locally defined default variations.
3. Spatial allocation for each relevant fuels and heating sub-categories with gridding based on information that distinguishes residential fuel use intensities on building-by-building basis (e.g. gridding based on GIS data on heating/cooling technologies in use and/or energy efficiency of buildings or city level building heating/cooling model with GIS capabilities). Housing and/or zoning modelling approaches are desirable to assess spatial changes in future projections. Temporal variation based on locally observed data.

Centralized energy production and industrial plants can often be dealt with as point sources, i.e. attain both location and activity and relevant technology data directly from the individual plant (level 3). However, sometimes such plant data are not available, and the spatial assessment of activities/technologies must be based on a surrogate type of approach. This means that the classification of complexity may again follow the three levels outlined above.

For **agriculture**, the requirements for its spatial resolution are not as high as for urban emission sources. Horizontal resolution of approx. 10×10 km^2 is often practical. In case detailed farm registers are available, activity estimates farm-by-farm basis (bottom-up) might be possible. However, at national level assessments, top-down allocation based on agricultural field areas or animal numbers might be sufficient.

Output

The output of the DRIVERS block is used as an input to PRESSURES. Therefore it needs to contain all relevant activity information for emission calculation. Activities used in the emission calculation typically include fuel use amounts, production intensities and kilometres driven aggregated in such a way to include emission-wise relevant classification of sectors, sources and technologies. Technological changes over time are important parameters for emission calculation, and are taken into account in the PRESSURE block. Especially for city level assessments, spatial patterns of activities and their change over time are essential.

Synergies among scales

Activity changes in the form of fuel switching and industrial production changes are affected largely at international (e.g. global markets) and national (e.g. national taxation) scale. On the other hand, population, housing and transport demand changes are affected largely at city (e.g. city taxation policies, general "attractiveness" of the city) and sub-city (e.g. traffic planning, zoning policies) scales.

Technological changes that are mainly of interest for the PRESSURE block are also affected at different scales. Many of the emission-related (e.g. traffic EURO standards, IE Directive) and climate-related (e.g. RE Directive) legislations that influence technological developments are defined at EU level. National level decisions may have a great impact as well (e.g. consumption or emission based vehicle taxation). At city level, it is possible to influence local problem spots (e.g. low emission zones, prohibitions of residential wood combustion) and set more general goals (city climate strategies) that influence technological developments.

Uncertainty

A short summary of the main challenges for the above emission source sectors is given in the following.

- Road traffic: Traffic models and/or detailed road segment specific traffic information are relatively commonly available. Technological parameters are relatively well known at least at national level. Parameters required for reliable non-exhaust emission assessment (e.g. road surface type and condition) can be a considerable source of uncertainty.
- Non-road traffic and machinery: For some forms of non-road activities, e.g. sea vessels, trains and airplanes, activities and spatial patterns are often relatively well known. For many other forms of machinery, in contrast, the activity data can be much more uncertain.
- Residential combustion: Residential wood combustion activities and technology information are often uncertain because a lot of the wood fuel is used from

private stock directly, and household level heating system stock is poorly known. Furthermore, spatial assessment (i.e. gridding) of residential combustion activities is often uncertain because of the lack of building registers for residential heating appliances.

2.3.2 *Pressures*

Air pollutant emissions act as pressures on the environment. Thus, the block PRESSURES of the IAM corresponds to the computation of the quantity of pollutants emitted into the atmosphere from stationary sources (such as smokestacks), surface areas (commercial or industrial facilities), and mobile sources (for example, road vehicles, locomotives, aircrafts, ships, etc.). The emission of a pollutant can in general be measured (as in large point sources) or estimated. These are generally calculated as the product of the activity of the emitter times an emission factor, that is the quantity of pollutant emitted per unit of activity.

Other possible pressures that affect air pollution concentrations are related to change of urban structures (new buildings, parks, etc.) that can modify the dispersion of the pollutants and so the concentrations. Similarly, strategies to mitigate Urban Heat Island (white or green roofs, etc.) may also have an impact on concentrations without modifying the emissions. These structural modifications in the city-level emission patterns are relevant, but at the moment very complex to be incorporated into a IAM scheme, and so will not be considered in the following descriptions.

Input

An emission is computed for a specific pollutant, an emission source, a spatial and temporal resolution. An emission inventory is a database combining emissions with a specific geographical area and time period (usually yearly-based) containing:

- The activity of the emission sources. For instance: the volume and the type of fuel burned, the number of kilometres travelled by the vehicles, etc. The activity data could be derived from (economic) statistics, including energy statistics and balances, economic production rates, population data, etc.;
- The amount of pollutant emitted by these sources per unit of activity, i.e. the emission factors.

The emission inventory may have different level of details depending on the availability of the data and their uncertainties. Data could be given per each activity sector, technology and fuel. For application of IAMs, information on costs and rates of application of technologies has to be integrated (normally with the assumption that costs remain linear with respect to rates of application).

The methodology used to estimate emissions depends on the objective of the study, the availability of the data and their uncertainty. In case of lack of detailed activity data or/and emission factors, it is necessary to collect such data at higher levels (national socioeconomic statistics, for example) to allow indirect calculations/estimations of the emission sources (Ponche 2002). Two main types of approaches are again distinguished:

- The top-down approach: used when, for a given area, there is lack of detailed data and to obtain the required emission resolution (scale) it is necessary to disaggregate the emissions calculated for a larger area. This approach computes the total amount of aggregated emission using for example data like total fuel consumption for the whole city or the whole country during a full year. This total is then distributed in time and space using the distribution of parameters linked with the activity responsible of the emissions (like population, road network, etc.).
- The bottom-up approach: used when for a given area numerous data at small scales can be collected and must be aggregated to higher sales. In the bottom-up approach, the emissions are directly computed from activity values in time and space.

The level of aggregation of the input data needed to apply these two types of methods is different. Usually, the bottom-up approach is preferred and also recommended to develop spatialized emission inventories (SEIs) and can reduce uncertainties. Nevertheless, the top-down approach is also generally used to control and correct the emission estimates. Applications show that in most cases the top-down and bottom-up approaches do not give the same results.

In order to harmonize European emission inventories, EMEP/EEA (2009a, b) proposed a guidebook with basic principles on how to construct an emissions inventory, the specific estimation methods and emission factors. In this guidebook, one key issue is the classification of the emission sources.

Classification of Emission Sources

The emission sources are usually at first classified in two classes depending on the emission process: natural sources and anthropogenic sources. They are also classified in three categories depending on their geographic characteristics, location and type:

- point sources, that are precisely located and often concern industrial sites, where large amount of atmospheric pollutant are emitted from very a small area (compared to the space resolution of the emission inventory);
- line sources, that correspond to main transportation infrastructures. If the traffic (road, air, railway, ship) on these routes is dense enough (relatively to the time and space resolutions of the emission inventory), they can be considered as continuous emission lines;
- area sources, that include all other sources as residential areas, industrial areas, etc., where numerous small emitters are spread/diffused.

In order to categorize the anthropogenic sources, several classifications in terms of activity, sectors and fuel use were proposed. At European level, SNAP97 (Selected Nomenclature for Air Pollution) is a reference classification proposed by

EEA, while in the EMEP/EEA (2009a, b) guidebook, NFR (Nomenclature for Reporting) classification developed under the Convention on Long-range Transboundary Air Pollution is used. This classification is completed by the list NAPFUE (Nomenclature for Air Pollution of FUEls), which allows to take into account all kinds of fuels used in the emission processes. For specific national, regional or local circumstances or needs, activities may be detailed based on more resolved categories. To help this work with the SNAP classification, EMEP/EEA (2009a, b) proposes a methodology to identify the major pollutants involved from all anthropogenic and natural emission processes. This handbook of default emission factors is especially useful in case of lack of specific knowledge of the processes used in the investigation area.

Spatialized Emissions Inventories (SEIs), Scenarios and Projections
Emission inventories are usually spatialized on a regular grid: the result is called spatialized emission inventory (SEI). The resulting SEI is used as input in the AQ part of an IAM to simulate the STATE, and is generally used as basis to simulate emission scenarios and projections.

Emission scenarios could be produced in several ways (EMEP/EEA 2009a, b) depending of the objectives of the studies:

- By modifying the activity index or data, as described in DRIVERS section.
- By modifying the emission factors of the emission generation processes. This includes new technologies or technological improvement, industrial processes, changes in fuel types or characteristics, energy saving (in terms of efficiency), composition of the vehicle fleet, etc.

The level of detail of the scenario is highly dependent on the level of classification of the sources and the data available for each category: in other words, the emission scenarios may be very simple and derived from the application of an emission reduction rate directly on the SEI; or they may be the results of assumptions on the future projections of the activities and the emission factors. Future emission factors should reflect technological advances, environmental regulations, deterioration in operating conditions and any expected changes in fuel formulations.

Functionality
The functionality of the PRESSURES box of an IAM aims at producing emission data or/and emission projections. The PRESSURES can be estimated through three different levels of complexity, depending on their further uses and the available data:

- **LEVEL 1**: emissions are estimated for rough sectors on a coarse grid (spatialization), using a top-down methodology. Uncertainties are not necessarily estimated. This level does not allow to perform detailed emissions projections.
- **LEVEL 2**: a combination of bottom-up and top-down methodology is used to calculate the emissions with the SNAP—NAPFUE classifications. Emissions factors and activity data representative of the area of study are used when available. Uncertainties are not necessarily estimated.

- **LEVEL 3**: emissions are calculated with the finest space and time resolution available, with the bottom-up method with all the SNAP-NAPFUE classifications details. Emission factors and activity data have to correspond to the specific activities of the studied area. The processes have to be detailed so that it is possible to attribute the most representative emissions. In case of lack of data, the top-down approach can be used but with the help of complementary data to take into account the regional specificities. The uncertainties may be quantitatively calculated, e.g. by a Monte Carlo method, whenever possible. This level is the best one to allow the generation of all kinds of scenarios provided that the emission changes are higher enough compared to the uncertainties of the SEI emission values.

Emission scenarios may be built directly from the SEIs by reducing the total emissions per grid box. These scenarios are then used in the STATE block to give general indications of the possible evolution of the air quality, or identify simplified equations that represent the links between emissions and concentrations in a complex IAM.

EMEP/EEA (2009a, b) classifies the methodologies to compute the emission projections:

- **LEVEL 1** projection methods can be applied to non-key categories and sources not expected to be modified by future measures. Level 1 projections will only assume generic or zero growth rates and simply projected or latest year's historic emission factors.
- **LEVEL 2** projections would be expected to take account of future activity changes for the sector, based on national activity projections and, where appropriate, take into account future changes in emission factors. It is necessary to have a detailed description of the source category in order to apply the appropriate new technologies or control factors to sub-sectors.
- **LEVEL 3** projections use detailed models to provide emission projections, considering additional variables and parameters. However, these models have to use input data that are consistent with national economic, energy and activity projections used elsewhere in the projected emissions estimates.

Output

A first output is an emission inventory that gives the total amount of different pollutants released into the atmosphere by all the different sources. These sources are classified using the processes producing the pollution (biogenic, industrial, transport-related, agricultural, etc.) and their type and spatial characteristics and distribution: point sources (industries, power plants, etc.), line sources (road transport) and area sources (biogenic, diffuse industries, residential areas, and small road sources).

A second output is a SEI that represents the amount of different pollutants released in each cell of a mesh. To get this SEI, the spatial information about the distribution of the sources (point, line and area) has to be projected on the mesh (normally a matrix of square cells). Then, the contribution of each source category for each pollutant is simply added. On the one hand, this resulting SEI can directly be used by an air pollution model. But, on the other hand, some information concerning the distribution of source categories as well as the accuracy of the source locations may be lost.

Synergies among scales
In theory, it is possible to use the spatial characteristics and locations of the emission sources in order to project the data on any kind of grid domain.

In practice, it is very difficult to manage, or even to find, a detailed and complete description of all the sources over large areas (scale of a continent or large countries). It follows that the first output of the large scale SEIs is based more on area than point and line sources in comparison to small scale SEIs. The sources of large scale SEIs are calculated using more top-down than bottom-up approaches. Consequently, the locations of the sources in large scale SEIs are not accurate and the projections of such SEIs on fine resolution grid lead to an overestimation of the sources dilution. It becomes then necessary to "re-concentrate" the sources using different earth surface characteristics defined at smaller scale. For example, the emission can be redistributed according to the land use (emissions release over the ground only and no emissions over water surfaces), the density of population (more emissions over dense population areas), the road network (road transport emissions only in cells crossed by roads), etc. Apart from simple redistribution proportional to these supplementary characteristics, which is typically done using linear regression, also more advanced approaches can be applied, e.g. using geostatistical methods, like kriging (Singh et al. 2011).

When using AQ models, it often happens that an accurate detailed emission inventory is available only on a part of the grid domain on which the study has to be performed. It is therefore necessary to combine data provided by different scale SEIs. In this situation, the best procedure is, first, to project all the SEI outputs on the same grid (using "re-concentration" when necessary) and then, to keep on each cell the data provided by the most accurate SEI. Even if there is a risk of inconsistency between the different SEIs because they have been produced using different methodologies (top-down or bottom-up for example) this procedure is a good compromise between consistency and accuracy.

Uncertainty
The uncertainties associated to emissions inventories (Werner 2009) are directly related to accuracy. This accuracy can be split into two main contributions:

- Structural inaccuracy, which is due to the structure of the inventory;
- Inaccuracy on the input data (i.e. activity data, emission factors).

The structural accuracy estimates the inventory structure ability to calculate as precisely as possible the real emissions. This uncertainty can be split into three contributions: inaccuracy due to aggregations (the emissions are calculated on defined spatial and time scales that may lack the information on the emission processes or on the variability of the real emissions); incompleteness (an emission inventory may be inaccurate due to the absence of emission sources); inaccurate mathematical formulation and calculation errors (the mathematical formulation used is generally highly simplified, and assumes, for example, that the relation between emission and activity is linear).

The uncertainties on the input data are mainly due to the lack of information on the different parameters used to estimate the emissions of an inventory. These emissions result mainly in the combination of input data like activity values and emission factors. The uncertainty on the values of input data can be due to simplification hypotheses, for example in the case of a large number of similar sources, supposed to have an average behaviour. They can be divided into four categories: extrapolation errors (when lacking emission factors or specific data related to some emissions sources, the corresponding values are extrapolated from other available data); measurement errors (they can lead to inaccurate activity data or emission factors); errors of copy (errors made during the reporting of values); errors in case of unknown evolution (future emission scenarios are associated to probability factors which can be seen as uncertainty or indefiniteness).

It is obvious that some relations exist between these different types of uncertainties and it is sometimes difficult to distinguish them.

The uncertainties of an emission inventory can be evaluated in a qualitative or quantitative way. The qualitative evaluation is mainly performed by experts (IPCC 2000; EPA 1996), while the quantitative one is based on error propagation methods and Monte Carlo methods. There is also a semi-quantitative method that can be used to evaluate the uncertainties, which consists in the rating of the data quality. Some numerical or alphabetical scores are attributed by experts to emission factors and activity data to describe the uncertainties of these data. There are two main classifications for these methods (see: EPA 1996): (1) the DARS method (Data Attribute Rating System) that attributes to each dataset a score ranging between 1 and 10 (the most accurate); (2) the AP-42 emission factor rate system that is the main reference in the USA but only for emission factors evaluation. The scores range from A (most accurate) to E. Both methods attribute scores, which are general indications on the reliability and the robustness of the data.

2.3.3 State

In the DPSIR approach, STATE is defined as the “environmental conditions of a natural system”. In the case of air quality, it describes the ambient concentrations of targeted pollutant (in specific applications also pollutant’s deposition). AQ state can be described as gridded concentrations/depositions over the studied area, or as local concentrations/depositions on receptor sites, depending on the objectives of the IAM and on the available tools. In addition to the spatial dimension, the AQ state also has a temporal dimension, considering that a pollutant can be monitored/modelled with a temporal resolution of hours/days, etc. Once concentrations/depositions are evaluated in space and time with the different available approaches, AQ indicators are usually calculated, such as aggregation of the initial AQ data to provide the number of PM10 daily exceedances on a cell, the annual mean of NO_2 aggregated over a domain, etc.

In the following, we focus on concentrations only as a state indicator, but the content would be basically the same for deposition.

It can be noticed that sometimes the PRESSURES block may be seen as acting directly on the IMPACT block, if simplifying the scheme and assuming a direct relationship between emissions and effects, with no evaluation of the STATE conditions.

Input

In IAM, the AQ state is described as the joint responses to pressures, constituting driving forces on which society can act at the spatial scale of the study, and external conditions, such as meteorology and pollution coming from the larger scale. Depending on the method chosen to perform an IAM, these forcing can be treated explicitly (this is the case when using a numerical model including meteorological and boundary conditions data), or act implicitly on other data. In certain cases, when AQ models are used for state evaluation, AQ observations can also be considered as input data, when these are used for model validation, data assimilation, or as initial or boundary conditions for models.

Functionality

The different methods that can be used to evaluate the AQ state, i.e. pollutant concentrations, are summarized in Fig. 2.3 and will be described in the following paragraphs. In parallel to the method used to define pollutant concentrations, methods are also often defined to estimate the contribution of the different emissions to the concentration (source apportionment).

The STATE block three-level classification is as follows:

LEVEL 1: The simplest way to characterize AQ state is to use measurements taken routinely, or during a measurement campaign (together with a geostatistic interpolation method if the aim is to obtain a map of concentrations over a studied area). Some studies also use the strong and highly uncertain hypothesis that local concentrations are proportional to local emissions to estimate source contributions.

LEVEL 2: It is based on a characterization of the AQ state using one model, adapted to the studied spatial scale. This model should be validated over the studied area and should use emissions input data also adapted to this scale. Concentrations used as boundary conditions of the model can be either extrapolated from measurements or extracted from a larger scale model. Observed concentrations can be used to correct the model (data assimilation) at least for the reference year, often used as a starting point for IAM applications. If the IAM is a prospective study, aiming to evaluate future policy scenarios, a method could be used to correct the model. A possibility in this context is to estimate, through data assimilation (if observations are available), map of increments/bias (related to the base case) to be used to "correct" the concentrations of future alternative emission reduction scenarios. Another input to the model are meteorological data, which can be obtained from observations or from a meteorological model. Spatial and temporal

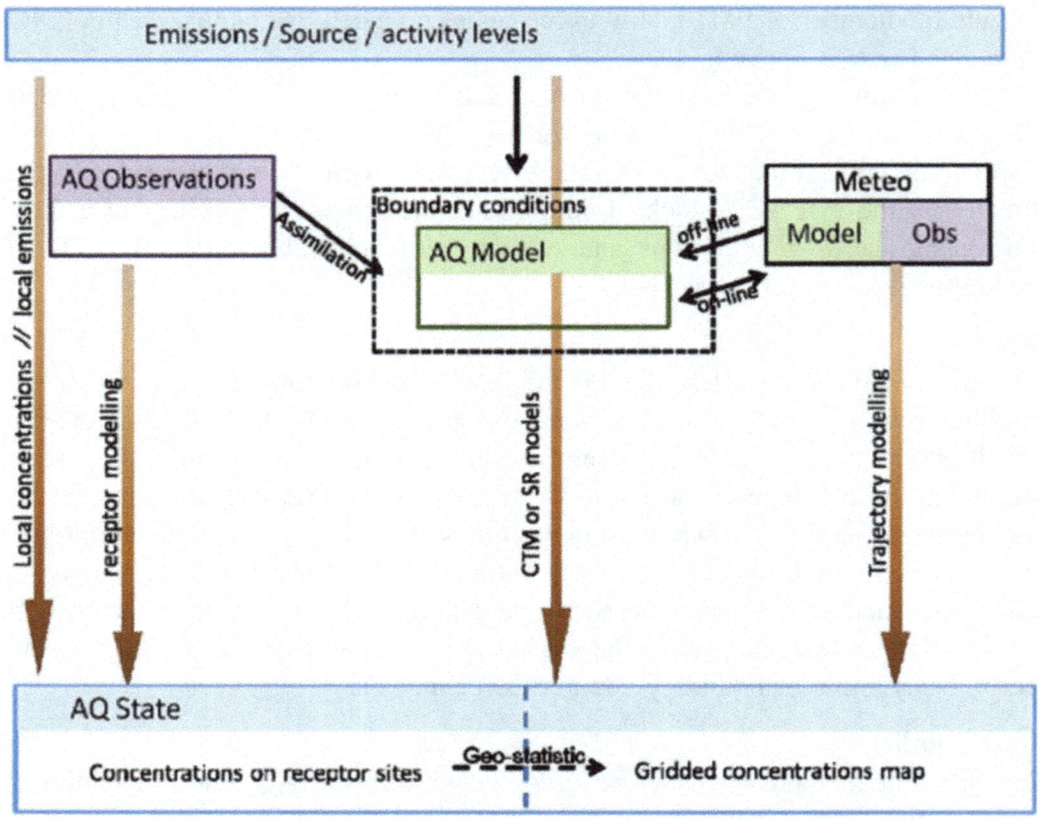

Fig. 2.3 Schematic of the different methodologies to estimate AQ state and to relate it to source contribution

resolution of the meteorological model should be adapted to that of the AQ model. For prospective IAM, using meteorological data from a specific year raises the problem of their representativeness, as it does not permit to catch the inter-annual variability of the meteorological conditions. To tackle this issue, one option could be to simulate more years, or in some way to "filter" the effect of the inter-annual variability in meteorology.

The full deterministic AQ model can be used to estimate contribution of the main sources on each grid point concentration, for example by cutting-off these sources one at a time. This method assumes the possibility of "adding" effects in some way and is time-consuming, as one full model run has to be done for each estimation of source contributions. Therefore, such calculations are generally limited to estimate large emission contribution over an area (e.g., industry, traffic, etc.). For some RESPONSES module implementations (as in the case of optimization approaches) thousands of model runs would be required, for example to minimize the cost of emission reduction measures. In such cases, the AQ model may be substituted by a more computational efficient source/receptor model (also called *surrogate model* or *meta-model*) based on simplifications of the AQ model. This model directly links the activity levels or the emissions to an AQ index calculated from targeted pollutant concentrations. The level of complexity of the surrogate model depends on the objectives of the IAM, on the nature of the pollutant (non-linearities, chemical reactivity, etc.) and, above all, on the output necessary for the subsequent IMPACT block (Carnevale et al. 2012b).

LEVEL 3: is based on a characterization of the AQ state using a downscaling models chain, both in term of AQ and meteorological models, from large scale (Europe, for example) to regional (country or regions) and local scale (city or street level). Using a downscaling model chain allows to take into consideration interactions between the various scales, such as transport of pollutant from large scale or interactions between mesoscale wind flows and local dynamics. Nesting between models can be one-way or two-ways, allowing local information to be passed to the larger scale model run. Sub-grid modelling approaches can also be used to combine different scales. The same model could be used for different parts of the chain, running the model itself at different resolutions; or different models could be applied at different scales, as local models (Gaussian models, for example) may use boundary conditions from a larger scale Eulerian model. Data assimilation and meteorological data representativeness issues are similar to those described for Level 2.

Output

The output of the STATE block may go from spatially and temporally-resolved concentrations of the targeted pollutants, i.e. hourly/daily concentrations on receptor sites or in each grid of the studied domain, to aggregated AQ indexes calculated through spatial/temporal aggregations. Typical aggregated indexes are, for instance, the number of PM10 daily exceedances, or annual mean of NO_2 in few or all domain cells. Other variable describing the STATE could be related to pollution depositions and climate change indicators (CO_2 emissions, global warming potential, etc.). In general, the choice of the correct output is based, on one side, on those adopted by the EEA, on the other, on their use for the calculation of IMPACT.

Synergies among scales

Using a downscaling model chain allows to take into consideration the interactions between different scales, both in terms of pollutant transport from large scale and in term of interactions between dynamic flows at various scale.

There is a close connection between climate change and air quality. Pollutant concentrations in the air are strongly influenced by changes in the weather (e.g., heat waves or droughts). At the same time, concentrations of pollutants such as O_3 and particles impact the climate through direct and indirect forcing. The first relation can be taken into account by using meteorological conditions from a climate model. However the relevance of using future climate meteorological conditions for short term studies (e.g., five years as in some cases in AQ plans) has not been demonstrated yet, as future meteorological conditions may not vary enough in 5 to 10 years. On the other way, estimating the impact of local changes in O_3 and particles on climate would require the use of meteorology-atmospheric chemistry coupled models at the regional scale. In this case, the STATE would not be the pollutant concentrations, but rather climate change related metrics, such as global warming potential or radiative forcing.

Uncertainty

When the AQ state is evaluated through measurements only, uncertainties are related to the measurements themselves, to the geostatistical methods used to interpolate point measurements and to the representativeness of measurement sites to characterize the area under study.

Uncertainties related to AQ numerical modelling have been widely discussed in the scientific literature. Intrinsic uncertainties of AQ modelling are mainly related to errors in the physical formulation of the model, and to uncertainties in the input data. An operational validation of the AQ model by comparison with measurements is required, opening the question of the representativeness of the chosen measurement sites in relation to the model scale. Evaluating the indefiniteness of prospective study is more challenging and would require the use of diagnostic evaluation (e.g., sensitivity tests) or probabilistic evaluation (e.g., errors propagation). Furthermore, as mentioned earlier, for prospective IAMs, estimating the AQ state over a relatively short temporal period (up to one year) introduces uncertainties on the representativeness of the estimated state itself.

2.3.4 Impact

The IMPACT block describes the consequences of any alterations or modifications of environmental conditions, being either beneficial or adverse. Among the various impacts, we could distinguish between impacts on human health, on environment (vegetation and ecosystems), on social, economic aspects or on climate. Moreover some impact could be derived from another, such as economic consequences of human health or of ecosystem services changes.

The choice of IMPACT would primarily allow to support the selection of the RESPONSES that would eventually influence the complete DPSIR chain.

Special attention will be paid in the following to health issues, that are important for local and regional decision making and are, in many cases, the most relevant impact from the economic viewpoint.

Input

Human health is a response to the exposure to a given air quality (STATE), and can be calculated using data that describe the air quality (such as level of concentration measured at a monitoring site, levels of concentration averaged for several monitoring stations or determined using an AQ model) and dose-response functions or concentration-response functions when available. In some case, the health impact can be calculated using data such as intake fractions computed after modelling the emissions to take into consideration (PRESSURES).

The choice of a pollutant to perform HIA (Health Impact Assessment) is often more restricted by the available knowledge on health effects and on the way to measure those effects, than by the input provided by the STATE block. The selection of input data depends in fact on the availability of a causal function to derive health output. The level of needed details on the exposure data depends on the output chosen, its occurrence and the strength of the causal relationship. However in general, the following input are needed to compute impacts:

- Air pollution concentrations
- Population data
- Dose-response functions.

Functionality

The input-functionality-output chain can be implemented at different levels of complexity. It depends on the strength and the robustness of the causal relationship between the exposure indicator (STATE or PRESSURES) and the health indicator chosen to support the decisions (RESPONSES) to be taken. The chosen approach to compute health impact (retrospective, prospective, counterfactual) does not restrict the level of complexity to be applied; it only demands more or less detailed data in the input-output chain.

- **LEVEL 1**: A coarse description of exposure provided either by measurement or modelling of AQ (e.g. average mean annual exposure for a city), a dose-response function or concentration-response function and a simple population description would give a rather coarse output. For examples: the number of hospital emergency visits related to increased ozone levels for a city or region.
- **LEVEL 2**: Similar to level 1, but with spatial details in the STATE description.
- **LEVEL 3**: A detailed temporal and spatial resolution for exposure and population data allows an accurate health analysis integrating, for instance, distance to roads, spatial distribution and vulnerable groups. For examples: The number of hospital emergency visits of those who live in greener or more trafficked areas of a city, related to local changes in ozone.

Output

The choice of health indicators to support decisions has to be made to show the potential policy action or inaction impact. Outputs have different strength in supporting policies. The burden of disease related to air quality can be expressed as such or translated into YOLL, DALY (Disability-Adjusted Life Year), life expectancy related to changes in exposure. Other indicators such as morbidity or mortality rate, number of hospital visits related to exposure and exposure changes can be used with a known dose-response or concentration-response function. The output representativeness strongly depends on the level of detail of population data.

The temporal resolution is also of importance, decisions on short-term exposure or on long-term exposure should be addressed separately using related health data.

Synergies among scales

Concerning the IMPACT and specially those on human health, the scale is strictly related to the level of uncertainties. The challenges of synergies encountered in

STATE and PRESSURES blocks will be emphasized in IMPACT with some more uncertainties and robustness issues. The description of the population data and their level of details will limit the potential of synergies among scales. As an example: Local scale IAM on one city will not show the same impact values than a larger scale IAM. Increasing coherence can be reached in computing a multiscale IAM with re-distribution to each local city of their own data.

Uncertainty
As not everybody is affected in the same way by air quality exposure, the HIA presents large uncertainties. Dose-response functions or concentration-response functions are identified as the main source of uncertainty in IAM. Epidemiologists often report an underestimation of causality. Therefore the literature recommends to use the available most detailed exposure estimate in epidemiological studies (e.g. for pollutants with high spatial variability this can be based on personal activity-based modelling or personal dosimetry), to assess the health effects of air pollution.

2.3.5 Responses

The RESPONSES block represent the Decision Framework, that is to say the set of techniques/approaches that are used to take decisions on emission reduction measures, or on activity changes, or on direct concentration reductions.

Input
Input required for this block may be:

- *Emissions.* They constitute the block input in those cases that do not use an explicit calculation of the STATE and of the IMPACT. Their spatial domain, discretization, and composition detail must be coherent with the detail of the possible actions;
- *Air Quality Indexes* (AQIs). Evolving pollutant concentration at different sites (measured or produced by some model) can be summarized into one or more AQIs. This often happens when an evaluation of the IMPACT is not performed. These AQIs are directly compared and/or combined in the RESPONSES block.
- *Impact.* This is the case when the full chain is implemented. While AQIs and impacts can be computed from measured data, to support decisions it is essential to compute them through (deterministic or statistical) models, since their variation has to be linked to possible actions.

As external forcing of the RESPONSES block, one has mainly to consider the decision setting in which the IAM will be used. This means that the range of actions that the local/regional authority can consider is clearly defined and the connection with other plans/regulations are explicit.

Functionality

The functionality of this block must suggest responses to the decision maker, to reduce precursor emissions (PRESSURES), or modify the DRIVERS, or directly act to improve the STATE (Vlachokostas et al. 2009).

The main components of this block are:

- *Control variables*: these represent the measures that can be applied by the regional/local Authority. They can be related to a macrosector or a pollutant level reduction (aggregated approach), or to a single technology acting on one or more pollutants (detailed approach). A further classification distinguishes between "end-of-pipe measures" (applied to reduce emissions at the "pipe" of an emitting activity) and "efficiency measures" (often called "non-technical measures", that reduce activity levels, e.g. acting on people behaviour, etc.).
- *Objectives*: these represent what a Decision Maker would like to improve/optimize. For instance, an objective could be to reach a given level of an AQI at minimum cost, or to use a predefined budget to minimize an AQI. More than one objective can be considered within the same problem (e.g. reducing two pollutants with a given budget).
- *Constraints*: these can be of different types, as legislative (i.e. new obligations on emission sources), economic (i.e. limited budget to be spent), physical (i.e. due to domain features), etc. Constraints can be mathematically formalized, if using a formal approach to take decisions; or they can be taken into account when making decisions, but without explicitly modelling them.
- *Implementation technique*: this represent, from an operational point of view, how all the ingredients already described (control variables, objectives, constraints) are put together and processed, to suggest one or more solution(s) to the problem. In some cases, the implementation would simply mean an expert advice, in other cases, the use of some piece of software running a suitable optimization procedure.

The RESPONSES block can again be described by three levels of complexity:

- **LEVEL 1**: Expert judgment and Scenario analysis. In this case the selection of measures to be adopted is based on expert opinion, with/without modelling support to test the consequences of a predefined emission reduction scenario. In this context, the costs of the emission reduction actions can be evaluated as an output of the procedure (even if in many cases they are not considered).
- **LEVEL 2**: Source Apportionment and Scenario analysis. In this case, the most significant sources of emissions are derived through a formal approach; this then allows to select the measures that should be applied. Again, emission reduction costs, if any, are usually evaluated as a model output.
- **LEVEL 3**: Optimization. In this case the whole decision framework is described through a mathematical approach (Carlson et al. 2004), and costs are usually taken into account. Different approaches (both in discrete and continuous setting) are available, as:

- *Cost-benefit analysis*: all costs (from emission reduction technologies to efficiency measures) and benefits (improvements of health or environmental quality conditions) associated to an emission scenario are evaluated in monetary terms and an algorithm searches for solutions that maximize the difference between benefits and costs among different scenarios.
- *Cost-effectiveness analysis*: due to the fact that quantifying benefits of non-material issues is strongly affected by subjective evaluations, the cost-effectiveness approach can be used instead. It searches for the best solutions considering non-monetizable issues (typically, health related matters) as constraints of a mathematical problem, the objective of which is simply the minimization of the sum of (relevant) costs (Amann et al. 2011).
- *Multi-objective analysis*: it selects the efficient solutions, considering all the objectives of the problem explicitly in a vector objective function (e.g., one AQI and costs), thus determining the trade-offs and the possible conflicts among them (Guariso et al. 2004; Pisoni et al. 2009).

Output

The outputs of the decision framework are the measures to be implemented to change the connected blocks. There are different options to describe these responses, as:

- Macrosector level emission reductions: reductions are applied to all emissions (PRESSURES) belonging to a CORINAIR macrosector. This is a very aggregated approach, but can provide policy makers with some insight on how to prioritize the interventions and it is easy to implement (Carnevale et al. 2012a, b).
- "End-of-pipe technologies" also called "Technical measures", (e.g. filters applied to power plant emissions, to cars, etc.). These measures are applied to reduce emissions (PRESSURES) before being released in the atmosphere. They neither modify the driving forces of emissions nor change the composition of energy systems or agricultural activities.
- "Efficiency measures" (or "Non-technical measures") are those, that reduce anthropogenic DRIVERS. Such measures can be related to people behavioural changes (for instance, bicycle use instead of cars for personal mobility, temperature reduction in buildings) or to technologies that abate fuel consumption (use of high efficiency boilers, or of building thermal insulating coats, which reduce the overall energy demand). Localization decisions (e.g. building new industrial areas, or new highways) can also be considered as "efficiency measures".
- Direct pollution reduction measures. These act directly on STATE to reduce the pollution already in the environment. Planting some species of PM absorbing trees in urban environments or using coatings photocatalytically decomposing nitrogen oxides belong to these types of measures.

Synergies among scales
The main issue of this type is the fact that regional authorities have to decide actions constrained by "higher levels" decisions, i.e. coming from national or EU scale. In practical terms, this means that regional scale policies are constrained to consider the national/EU legislation as a starting point for their choices. In the effort to "go beyond CLE" within their regional domain, some "higher level" constraints cannot be disregarded or modified. This issue has to be considered for both Air Quality and Climate Change fields. In both cases, in fact, there are a lot of agreement/protocols that are already in force.

Uncertainty
As stated in UNECE (2002), it is important that the decisions focus on robust strategies, that is to say on "policies that do not significantly change due to changes in the uncertain model elements". This issue is linked to the need of defining a set of indexes and a methodology to measure the sensitivity of the decision problem solutions. It is in fact worth underlining that, while for air quality models the sensitivity can be measured by referring in one way or the other to field data (Thunis et al. 2012), for IAMs this is not possible, since an absolute "optimal" policy is not known and most of the times does not even exist. The traditional concept of model accuracy must thus be replaced by notions such as risk of a certain decision or regret of choosing one policy instead of another.

Acknowledgments This chapter is partly taken from APPRAISAL Deliverable D3.2 (downloadable from the project website http://www.appraisal-fp7.eu/site/documentation/deliverables.html).

References

Amann M, Bertok I, Borken-Kleefeld J, Cofala J, Heyes C, Höglund-Isaksson L, Klimont Z, Nguyen B, Posch M, Rafaj P, Sandler R, Schöpp W, Wagner F, Winiwarter W (2011) Cost-effective control of air quality and greenhouse gases in Europe: Modeling and policy applications. Environ Model Softw 26:1489–150

Carnevale C, Finzi G, Pisoni E, Volta M, Wagner F (2012a) Defining a nonlinear control problem to reduce particulate matter population exposure. Atmos Environ 55:410–416

Carnevale C, Finzi G, Guariso G, Pisoni E, Volta M (2012b) Surrogate models to compute optimal air quality planning policies at a regional scale. Environ Model Softw 34:44–50

Carlson DA, Haurie A, Vial J-P, Zachary DS (2004) Large-scale convex optimization methods for air quality policy assessment. Automatica 40:385–395

EEA (2012) Europe's environment: an assessment of assessments. EEA, Copenhagen. Available from http://www.eea.europa.eu/publications/europes-environment-aoa. Last accessed March 2016

EMEP/EEA (2009a) Air pollutant emission inventory guidebook—2009, EMEP/EEA. Available from http://www.eea.europa.eu/publications/emep-eea-emission-inventory-guidebook-2009. Last accessed March 2016

EMEP/EEA (2009b) Air pollutant emission inventory guidebook

EPA (1996) Evaluating the uncertainty of emission estimates, EIIP emission inventory improvement program. Available from http://www.epa.gov/ttn/chief/eiip/techreport/volume06/vi04.pdf. Last accessed March 2016

Guariso G, Pirovano G, Volta M (2004) Multi-objective analysis of ground-level ozone concentration control. J Environ Manage 71:25–33

IPCC (2000) Good practice guidance and uncertainty management in National Greenhouse Gas Inventories. Available from http://www.ipcc-nggip.iges.or.jp/public/gp/gpgaum.htm

Nagl C, Moosmann L, Schneider J (2005) Assessment of plans and programmes reported under 1996/62/Ec—final report, service contract to the European Commission—DG Environment, Contract No. 070402/2005/421167/MAR/C1. Avaliable from http://ec.europa.eu/environment/air/quality/legislation/pdf/assessment_report.pdf. Last accessed March 2016

Pisoni E, Carnevale C, Volta M (2009) Multi-criteria analysis for PM10 planning. Atmos Environ 43:4833–4842

Ponche J-L (2002) The spatialized emission inventories: a tool for the Air Quality Management French IGPB-WCRP News Letters (in French)

Singh V, Carnevale C, Finzi G, Pisoni E, Volta M (2011) A co-kriging based approach to reconstruct air pollution maps, processing measurement station concentrations and deterministic model simulations. Environ Model Softw 26:778–786

Thunis P, Georgieva E, Pederzoli A (2012) A tool to evaluate air quality model performances in regulatory applications. Environ Model Softw 38:220–230

UNECE (2002) Progress report prepared by the chairman of the task force on integrated assessment modelling. United Nations Economic Commission for Europe, Geneva

Vlachokostas C, Achillas Ch, Moussiopoulos N, Hourdakis E, Tsilingiridis G, Ntziachristos L, Banias G, Stavrakakis N, Sidiropoulos C (2009) Decision support system for the evaluation of urban air pollution control options: Application for particulate pollution in Thessaloniki, Greece. Sci Total Environ 407:5937–5948

Werner S (2009) Optimization of the spatialized emission inventories: estimation of uncertainties, determination of the emission factors of black carbon from the road traffic. Application to the ESCOMPTE and Nord Pas-de-Calais Region emission inventories. PhD Thesis, University of Strasbourg, Department of Chemistry (in French)

Chapter 3
Current European AQ Planning at Regional and Local Scale

C. Belis, J. Baldasano, N. Blond, C. Bouland, J. Buekers, C. Carnevale, A. Cherubini, A. Clappier, E. De Saeger, J. Douros, G. Finzi, E. Fragkou, C. Gama, A. Graff, G. Guariso, S. Janssen, K. Juda-Rezler, N. Karvosenoja, G. Maffeis, A. Martilli, S. Mills, A.I. Miranda, N. Moussiopoulos, Z. Nahorski, E. Pisoni, J.-L. Ponche, M. Rasoloharimahefa, E. Real, M. Reizer, H. Relvas, D. Roncolato, M. Tainio, P. Thunis, P. Viaene, C. Vlachokostas, M. Volta and L. White

3.1 Introduction

This chapter provides a review, derived from the extended survey conducted within the APPRAISAL project, of the integrated assessment methodologies used in different countries to design air quality plans and to estimate the effects of emission abatement policy options on human health.

The final purpose of this review is to foster the dissemination of knowledge on integrated assessment for air quality planning at regional and local scales, and to

C. Belis · E. De Saeger · E. Pisoni · P. Thunis
European Commission, Joint Research Centre (JRC), Directorate for Energy, Transport and Climate, Air and Climate Unit, Ispra, Italy

J. Baldasano
Barcelona Supercomputing Center—Centro Nacional de Supercomputacion, Barcelona, Spain

N. Blond
Centre National de la Recherche Scientifique (CNRS), Paris, France

C. Bouland · M. Rasoloharimahefa
Université Libre de Bruxelles (ULB), Brussels, Belgium

J. Buekers · S. Janssen · P. Viaene
Vlaamse Instelling Voor Technologisch Onderzoek N.V. (VITO), Mol, Belgium

C. Carnevale · G. Finzi · M. Volta (✉)
Università Degli Studi di Brescia, Brescia, Italy
e-mail: marialuisa.volta@unibs.it

G. Guariso and M. Volta (eds.), *Air Quality Integrated Assessment*,
PoliMI SpringerBriefs, DOI 10.1007/978-3-319-33349-6_3

provide policy makers and regulatory bodies across EU member states with a broader understanding of the underlying scientific concepts.

The survey allowed to populate a structured database (http://www.appraisal-fp7.eu), designed in collaboration with experts involved in the design of Air Quality Plans (AQP), aimed at identifying methodologies adopted in Europe to define AQ plans. The following topics were considered: (1) synergies among national, regional and local approaches, including emission abatement policies; (2) air quality assessment, including modelling and measurements; (3) health impact assessment approaches; (4) source apportionment; and (5) uncertainty and robustness, including Quality Assurance/Quality Control (QA/QC).

The APPRAISAL database currently totals 59 contributions from 13 MS, fully checked for consistency and completeness. Though probably not being completely representative from the statistical viewpoint, they provide a good prospect on the current EU situation and clearly indicates some of the actual trends. Two groups of respondents were distinguished to refine the analysis: the stakeholders involved in the design of "air quality plans" (AQP) and groups involved in "research projects" (RP). While AQP, which represent 58 % of the database information coming from 10 MS, is representative of current practices in the decision process, RP (31 % of

A. Cherubini · G. Maffeis · D. Roncolato
TerrAria Srl, Milan, Italy

A. Clappier · J.-L. Ponche
Université de Strasbourg, Strasbourg, France

J. Douros · E. Fragkou · N. Moussiopoulos · C. Vlachokostas
Aristotle University of Thessaloniki, Thessaloniki, Greece

C. Gama · A.I. Miranda · H. Relvas
Universidade de Aveiro, Aveiro, Portugal

A. Graff
Umweltbundesamt, Dessau, Germany

G. Guariso
Politecnico di Milano, Milan, Italy

K. Juda-Rezler · Z. Nahorski · M. Reizer · M. Tainio
Systems Research Institute of the Polish Academy of Sciences (IBS-PAN),
Warsaw University of Technology, Warsaw, Poland

N. Karvosenoja
Suomen Ymparistokeskus (SYKE), Helsinki, Finland

A. Martilli
Centro de Investigaciones Energeticas,
Medioambientales y Tecnologicas (CIEMAT), Madrid, Spain

S. Mills · L. White
AERIS Europe Limited, West Sussex, UK

E. Real
Institut National de l'Environnement et des Risques (INERIS),
Verneuil-En-Halatte, France

the database contributions, coming form 7 MS) are usually assumed to be based on the most updated methods. Seven studies stored in the database are classified as 'Other'. Countries represent the study area in 20 % of cases, regions in 25 % and agglomeration or urban level in 30 % of the cases (the remaining percentage refers to other types of focus which could not be classified in these categories). The current status (September 2015) of the databases is presented in Fig. 3.1 where the contributions are shown per country. Local planning authorities (e.g. municipality) represent 25 % of the respondents whereas universities, research institutions, environmental agencies represent each, about 20 %.

In order to characterize the operational use of AQ assessment and planning modelling tools, the APPRAISAL questionnaire includes the following information for each air quality plan: the overall purpose of the activity (air quality assessment, mitigation and planning, source apportionment), the strategy followed (scenario analysis, cost-benefit, cost-effectiveness, multi-objective approach), the source/receptors (methodology, spatial and temporal resolutions, indicators), the modelling approaches (models, processes, spatial and temporal resolutions, nesting), the input data including emissions (inventory approach, split into activity sectors, resolution, etc.), meteorology (models, processes, time and spatial resolution), initial and boundary conditions. Also the use of measurements was investigated (measurements method, type and location of the monitoring stations, temporal resolution, transformation of the data if any).

In the following section, the DPSIR blocks used to describe the plans are analyzed. AQ plan scales and uncertainty, two common and transversal topics, are discussed in the second section, while a methodology to classify the Air Quality Plans in Europe is proposed in the last part of the chapter.

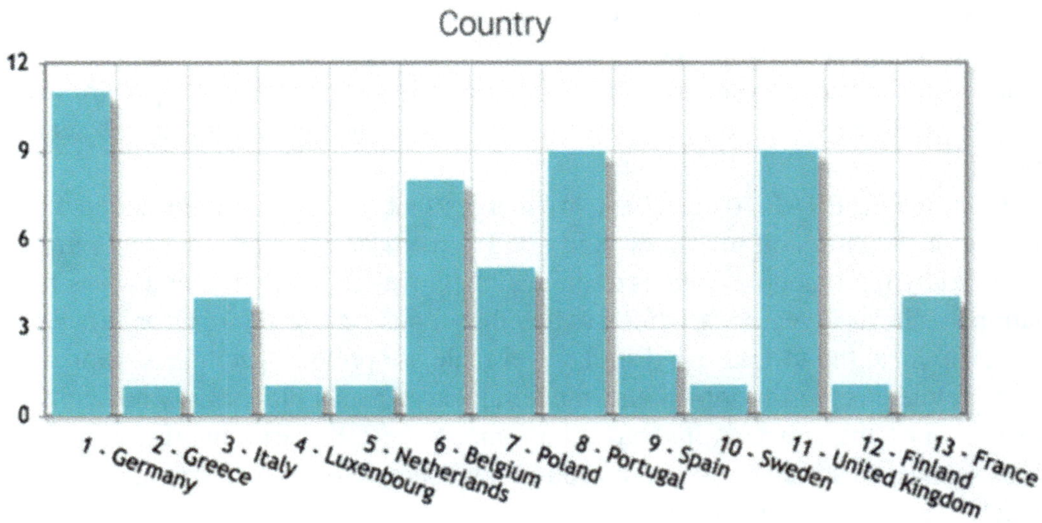

Fig. 3.1 Screenshot of the query to the online APPRAISAL database relative to the contributions in terms of countries

3.2 Actual Use of IA Components

This section focuses on the methodologies developed in recent years and implemented in the AQP and RP reported in the APPRAISAL database for each DPSIR component.

The database collects both AQPs and RPs. The rationale for this is that whilst AQPs are a consequence of air quality assessment and limit value exceedances actually detected, RPs might have a broader scope since there are no such formal constraints that have to be obeyed and so they may go beyond what is current practice.

3.2.1 Drivers and Pressures

The function of the DRIVERS block is to model the development of the driving activities (i.e. road traffic, off-road traffic and machinery, residential combustion, centralized energy production/industry, agriculture) over time. It is the direct and basic input to the PRESSURES block in the form of, e.g., road traffic kilometres driven, residential heating fuel consumption etc. The PRESSURE block holds the information on the quantity of pollutants emitted into the atmosphere from all the different sources. The emission of a pollutant can be measured or estimated. These are generally calculated as the product of the activity of this emitter and an emission factor, that is the quantity of pollutant emitted per unit of activity. There are also pressures affecting air pollution concentrations that are related to changes in urban structures (new buildings, roads, trees etc.) that can modify the dispersion of the pollutants.

At the moment, it is very complex to incorporate structural changes in a IAM scheme; so they have not been included in the current study.

Even though it is well known that emission inventories do not represent the actual contribution of sources to atmospheric pollution, many local governments use them directly as source identification tools for the design of abatement measures.

From the APPRAISAL database, it emerges that, in general, the scale and resolution of the emission inventory is in good agreement with the scale and purpose of the study (and model). Studies at the national level generally use emissions from national official inventories while studies that focus on the regional or urban (1–5 km), to local (up to 1 km) and street level scale use project specific emission data. In principle, the resolution of the modelling system should be in line with the resolution of the emission inventory but among the 59 questionnaires, 5 applications seemed to use an emission resolution not adapted to the geographical zone for which the study was intended.

Emissions are classified according to their sources. In the APPRAISAL questionnaire, the Selected Nomenclature for reporting of Air Pollutants (CORINAIR SNAP code) is used. This nomenclature was originally developed by the EEA's

European Topic Centre on Air Emissions (ETC/AE) and is common for emission inventories used as model inputs. In this nomenclature, sources are classified in three levels of details:

- Macro-sectors (SNAP level 1, e.g. "energy transformation sector"); it exists 11 different macro-sectors,
- Sectors activity (SNAP level 2, e.g." public power") which are a disaggregation of macro-sectors level,
- Activity levels (SNAP level 3, e.g. "combustion plants ≥300 MW (boilers)") which are a further disaggregation of sectors levels.

For each disaggregation level, more details can be added with definition of fuel specification.

Emission inventories with disaggregation to the sector activity and activity levels are most commonly used (Fig. 3.2). Together they cover one half of the questionnaires. Only 10 % of the studies use a macro-sector disaggregation level. A combination of different levels of disaggregation is often used. Fuels specification is used in more than 50 % of the cases. According to the database, there is no relation between the category disaggregation level and the spatial scale of the study.

Concerning the approach used to set up the inventory, a combined approach using both a bottom-up and top down methodology is most common (58 %). This is not surprising as official national and regional inventories are usually constructed using this complementary approach. A top-down approach alone is used in few cases (8 %), while bottom-up approaches alone represent about 22 % of the cases. For the studies using a bottom-up approach, a majority of them have created a project specific emission inventory over a small area. Urban, local and street level studies represent more than 80 % of the studies using a bottom-up approach.

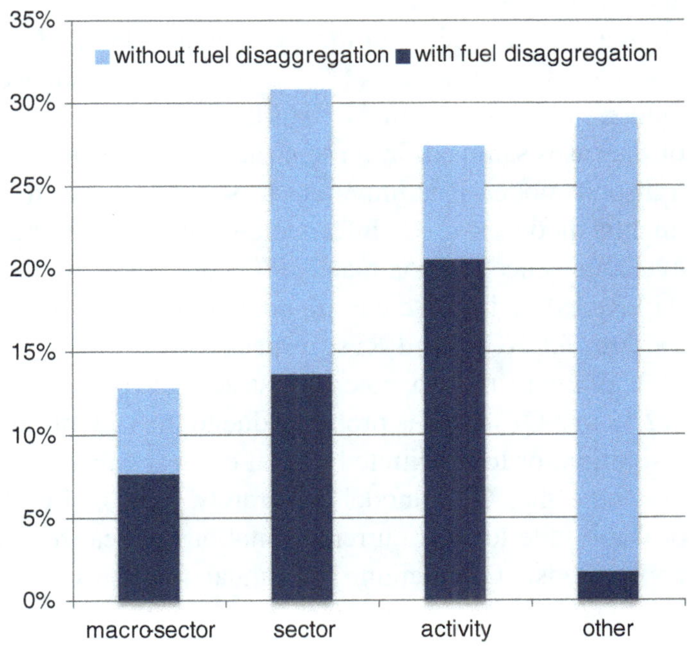

Fig. 3.2 Disaggregation level used in AQP and RP as reported in the APPRAISAL database

3.2.2 State

STATE, in the DPSIR approach, is defined as the environmental conditions of a natural system; in the current case, it represents the concentrations of targeted pollutant in atmosphere.

Air quality state can be described as gridded concentrations over the studied area, or as local concentrations at receptor sites. The AQ state has also a temporal dimension, considering that a pollutant can be monitored or modelled with different temporal resolutions.

A large variety of chemical transport models exist, implementing from simpler to more complex approaches and covering different scales, going from global/regional scales to urban and street level scales. State can be also described by source-/receptor models that directly link the emission to an AQ index calculated from targeted pollutant concentrations.

The APPRAISAL database indicates that national, regional and local authorities use a large variety of air quality models to design their AQPs and assess their impacts on air quality.

If we analyze the responses in terms of model types, Eulerian models are the most used with 32 and 59 % for AQP and RP, respectively (Fig. 3.3) which is not surprising since Eulerian modelling can be applied from the regional down to the local scale. In the case of AQPs, Gaussian plume and puff approaches represent about 20 %, in total while in RPs they represent only 6 % of use cases.

In total 33 different model names are mentioned. The most popular are the Eulerian models CAMx with 8 citations and CHIMERE with 11. CALPUFF is cited 6 times in the sample, but also traffic models are included (IMMIS, PROKAS and OSPM) with more than 5 citations. The many different models that are used today are a clear indication that no standard reference model currently exists. It is also interesting to note that in many AQPs, more than one model is used: three or more are used in 33 % of the cases, while about 27 % of the AQPs refer the use of two models and about 44 % of a single model. Regarding research projects, a unique model is used in 44 % of the cases, two in 17 % and three or more in 39 % of the cases sampled. In these projects, CHIMERE is the most often used chemical transport model. It is important to stress however that in one reported case, no air quality model is used. Information about modelling methodologies is in general available since approximately 70 and 85 % of the models referred to by the APPRAISAL database contributors are included in the EEA Model Documentation System, for AQPs and RPs, respectively.

It is interesting to note that street canyon models are not so frequently used (12 % in AQP). This is probably due to the lack of proper input data at the adequate resolution, or to the limited spatial coverage these models generally have. One can also note that CFD models are rarely used in Europe even in research projects, probably due to their current limitation to idealized, stationary and very fine scale applications. Calculation of annual statistics therefore still remains a very

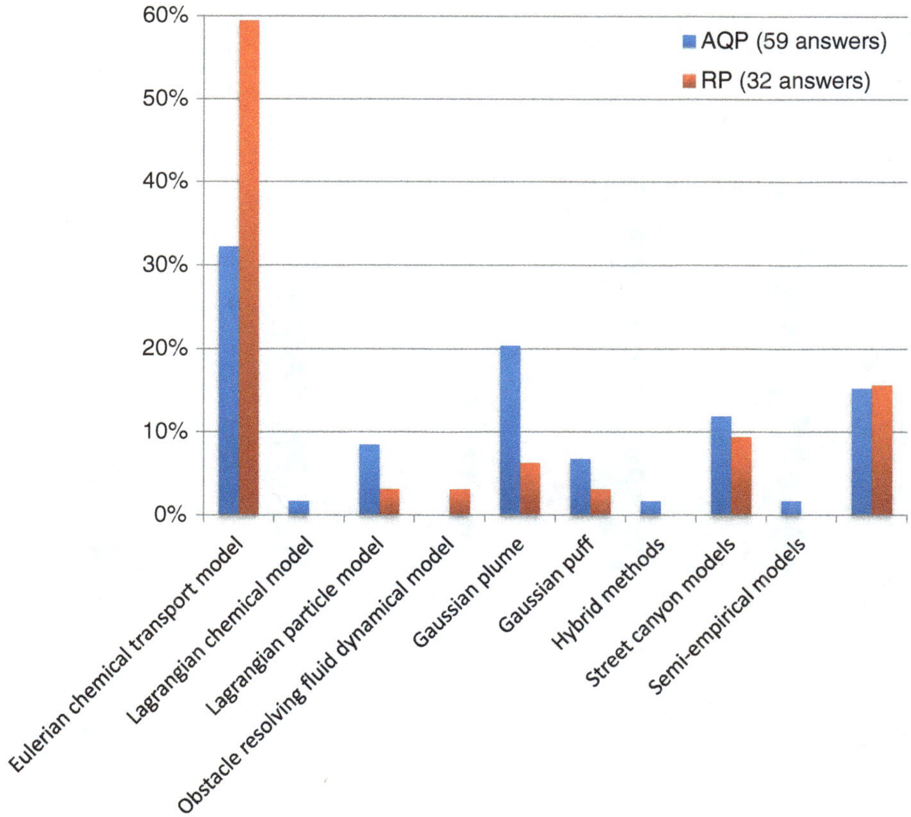

Fig. 3.3 Model types as used in AQP (*blue*) and RP (*red*)

challenging task for this type of models as shown in Parra et al. (2010), who attempted to estimate the concentration evolution from a series of steady state simulations for long time periods. With increasing computer power their importance might however increase in the future as they could progressively take on the role of the current generation of empirical or Gaussian models for local and street level modelling. The "hybrid" models in Fig. 3.4 refer to the application of a method based on numerical and statistical models.

The spatial scale of the AQ models was analyzed. Since at least 3–4 grid points are needed to resolve a flow structure, models with a resolution coarser than 3 km were classified as "regional scale" (5–50 km) while models with a resolution coarser than 500 m were considered as "urban scale" (1–5 km). The "local scale" (up to 1 km) models were those with a resolution between 500 and 10 m and finally "street scale" models are those with a resolution in the order of meters.

In total, the 59 air quality studies with up to 3 AQ models each, lead to a total of 177 different model setups (Fig. 3.4). Among RP studies, 40 % of AQ models were for the regional scale, 30 % at the urban scale, 13 % at the local scale, and 11 % at street scale. For the remaining 6 % model setups no information was given on resolution or range of scales.

Although the majority of the AQP applications regard regional and urban scales, exceedances of air quality limit values occur at traffic-induced hot spots to a large

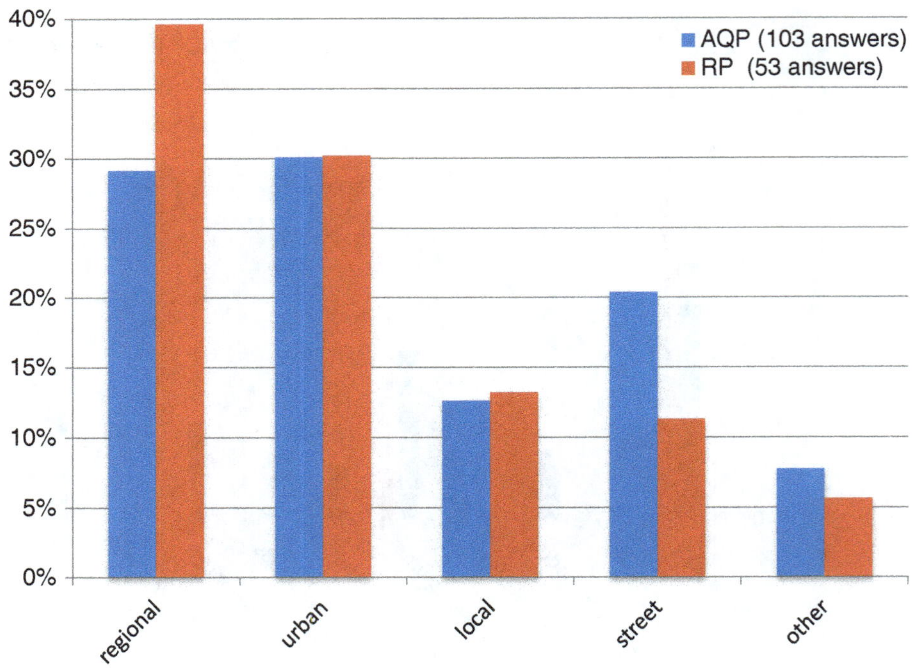

Fig. 3.4 Main scope for air quality assessment with respect to spatial scale for AQP (*blue*) and RP (*red*). Regional ranges from 10 to 50 km; urban from 1 to 5 km and local below 1 km

extent. Consequently, some of the AQPs adopt street level (20 %) and/or local scale modelling (13 %).

Only 12 % of the AQPs report on the use of highly resolved street canyon models. Even if alternatives to explicit street canyon modelling exist and consist in extending regional/local scale model capabilities to account for sub-grid scale effects, in the majority of the cases (more than 80 %) reported in the APPRAISAL survey, no additional model feature is included in the modelling approach to capture street effects, although these are keys to reproduce the concentrations and frequent exceedances at street locations.

More complex IAM methodologies, in which optimization algorithms are implemented, cannot embed full 3D deterministic multi-phase modelling systems for describing the nonlinear dynamics linking precursor emissions to air pollutant concentrations because of their computational requirements. They therefore rely on simplified relationships for describing the link between emissions and air quality, which are called source/receptor models (S/R).

In terms of the Design of Experiment required to identify these S/R, the majority of approaches apply the OaT (Once at a Time) approach (11 studies), in which one varies one emission at a time, and measures the variation in the concentration or effects at one site. In few cases "factor analysis" (2 studies) in which the impact of an emission and its interactions with another factor are considered simultaneously or a "statistical based" approach (3 studies), based on global sensitivity indexes, are used. The number of scenarios considered in the Design of Experiment is more than 50 (Fig. 3.5) only in the case of more complex research projects. The number of

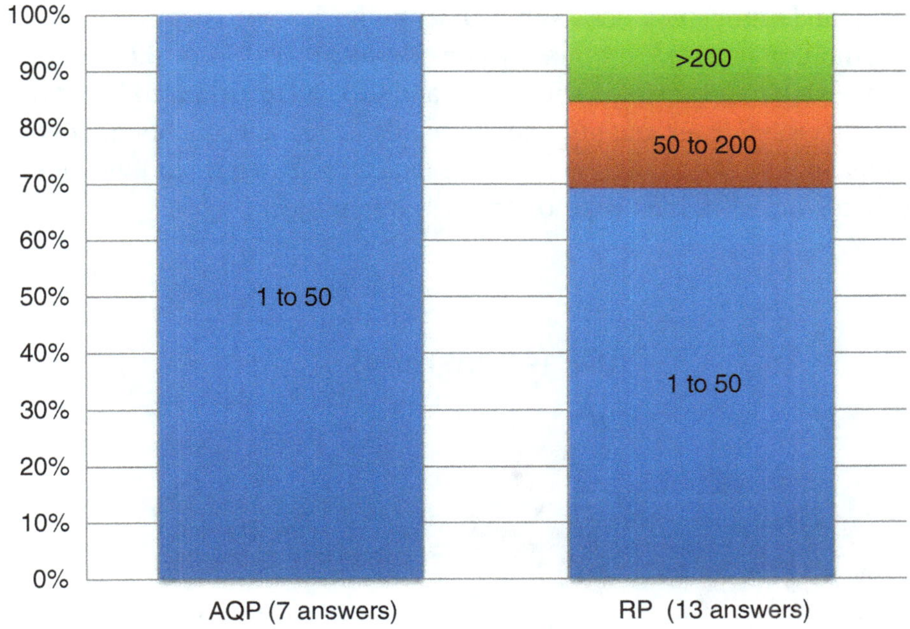

Fig. 3.5 Number of AQ modelling simulations (runs) used to identify S/R considered for AQPs and RPs

meteorological years considered is limited to a single year in 70 % of both the AQPs and RPs.

3.2.3 Impact

The block on IMPACT describes the consequences of modifications of environmental conditions related to the STATE of air quality, being either beneficial or adverse.

Only 34 studies included the assessment of Health Impact (HI): 21 Air Quality Action Plans, 11 Research Projects and 2 other activities. But, only 5 questionnaires have specifically expressed HIA as the main objective, respectively 4 for research projects and 1 for another activity. This reflects the fact that Integrated Assessment Models do not all necessarily include the health aspects and the AQPs are designed not with the main scope to assess HI.

The most common approaches used for HIA are the predictive approach (11 times) and the retrospective approach (7 times), while the counterfactual approach had been answered 2 times and other methods 14 times.

Among all the activities, 11 (6 of which were AQPs) HIAs focused on both short-time and long-term exposure to pollutants, 10 focused on long term exposure and 1 on short-term exposure.

The most frequent air pollutants included in the health impact assessments are related to the urban pollutants, such as particles (PM10 and PM2.5) followed by ozone (O_3) and nitrogen oxides (NOx). Other pollutants such heavy metals (arsenic, nickel, cadmium and lead) are mainly considered in RPs (Figs. 3.6 and 3.7).

The exposure indicators, for both AQPs and RPs, were estimated based on intake fraction (emissions), air quality monitored data and air quality modelled data (Figs. 3.8 and 3.9). Additionally, exposure indicators based on individual exposure data were also used in the scope of one research project.

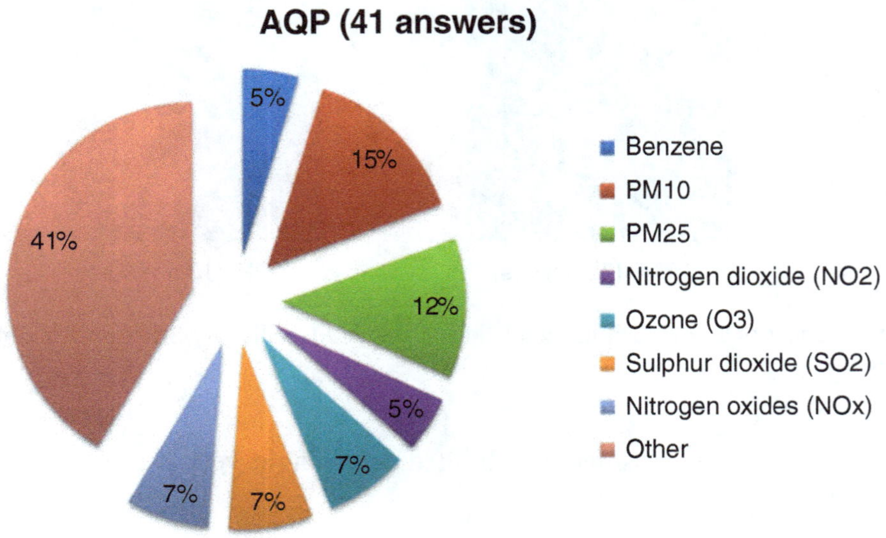

Fig. 3.6 Air pollutants assessed in HIA for AQPs

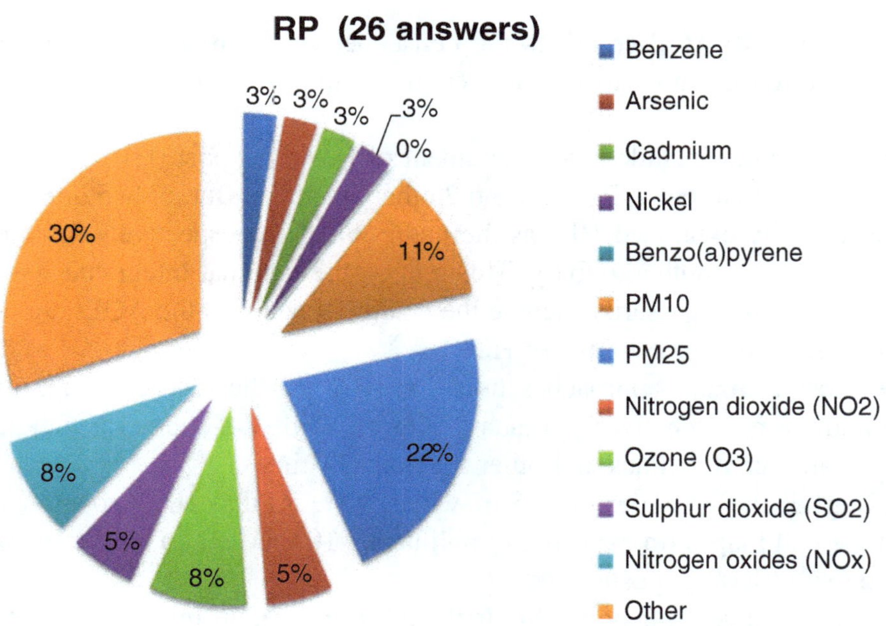

Fig. 3.7 Air pollutants assessed in HIA for RPs

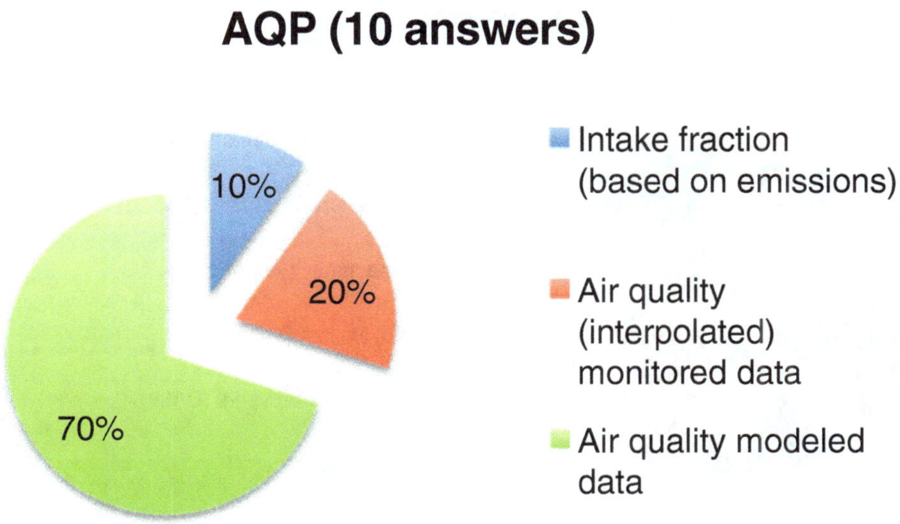

Fig. 3.8 Distribution of the calculation for exposure indicators in AQPs

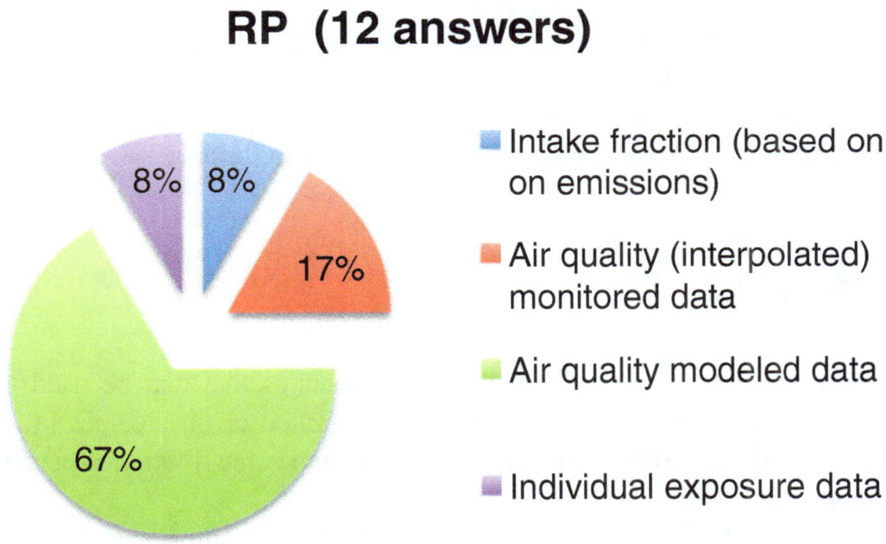

Fig. 3.9 Distribution of the calculation for exposure indicators in RPs

The spatial resolution considered for population and concentration estimation is usually the same. The temporal resolution used for concentration data differs between the two types of activities: 5 of the assessed AQPs use daily temporal resolution, 2 hourly and 2 annual. Six RPs utilize daily resolution, 2 are based on annual data and 2 on hourly data.

In the case where monitored concentration levels were used for the assessment of exposure, 2 studies processed data recorded at traffic station sites, 5 studies used data from urban background stations, 4 from sub-urban background sites and only one from rural background station.

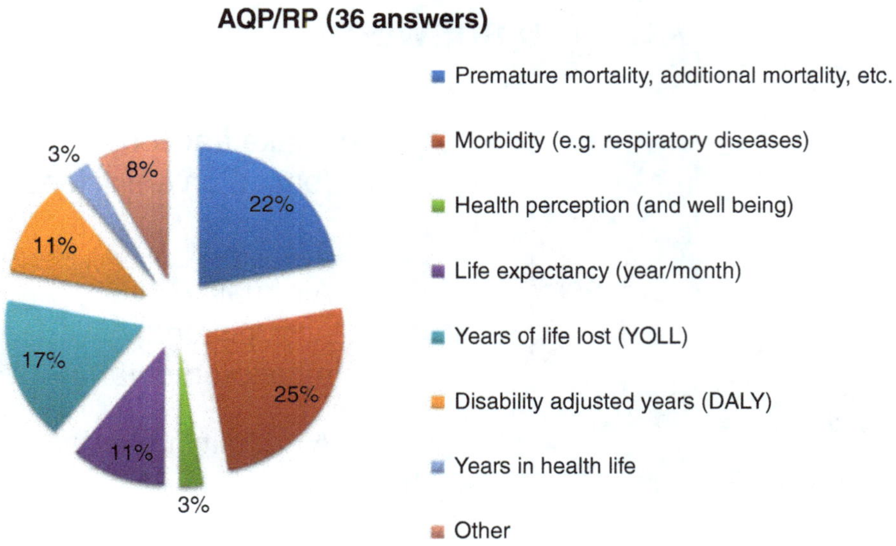

Fig. 3.10 Health indicators in the AQPs and RPs

Approximately 20 % of the AQPs that undertook a HIA considered a sub-group based on the age of the population. RPs also focused on the sub-groups gender and on other variables, beside age.

The considered HI indicators were related to premature mortality and morbidity (Fig. 3.10). Only two studies did not consider mortality impact.

3.2.4 *Responses*

This block represents the set of techniques/approaches that can be used to take decisions on emission reduction measures to be applied or on changes in activity levels (drivers). The DPSIR framework helps to visualize the difference between the possible approaches (Fig. 3.11).

All the items stored in the database implemented modelling systems to define mitigation measures and planning (Fig. 3.12). RPs are more oriented than AQPs to planning and source apportionment.

The Scenario analysis is the most frequently used methodology (Fig. 3.13), both in AQPs (more than 60 % of the cases) and RPs (roughly 30 % of the cases) implementation.

In the scenario analysis approach, source-apportionment can be used to identify the main emission sources that contribute to air pollution concentrations. Emission reduction measures are selected and/or established taking into consideration synergies at different scales. The effect of these measures on the air quality improvement is quantified using air quality modelling systems and afterwards translated to

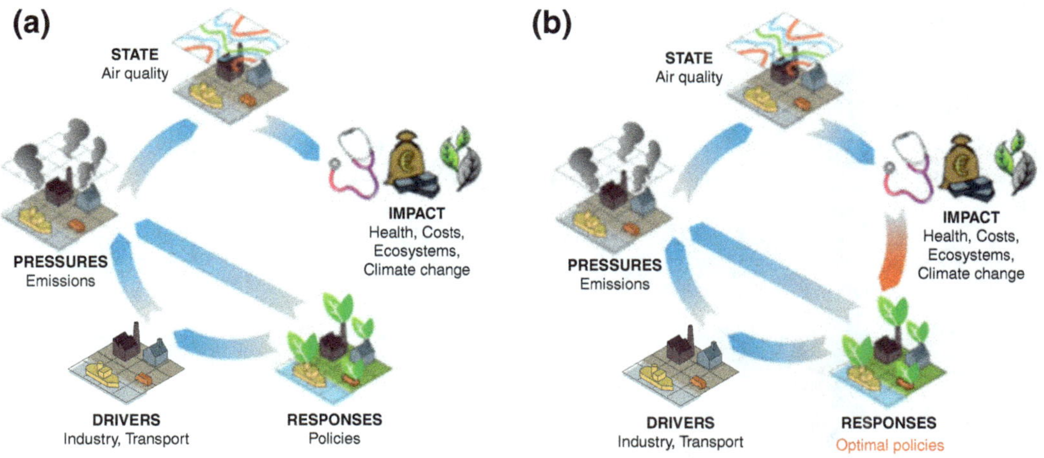

Fig. 3.11 IAM approaches within the DPSIR scheme: scenario analysis (*left*) and optimization approach (*right*)

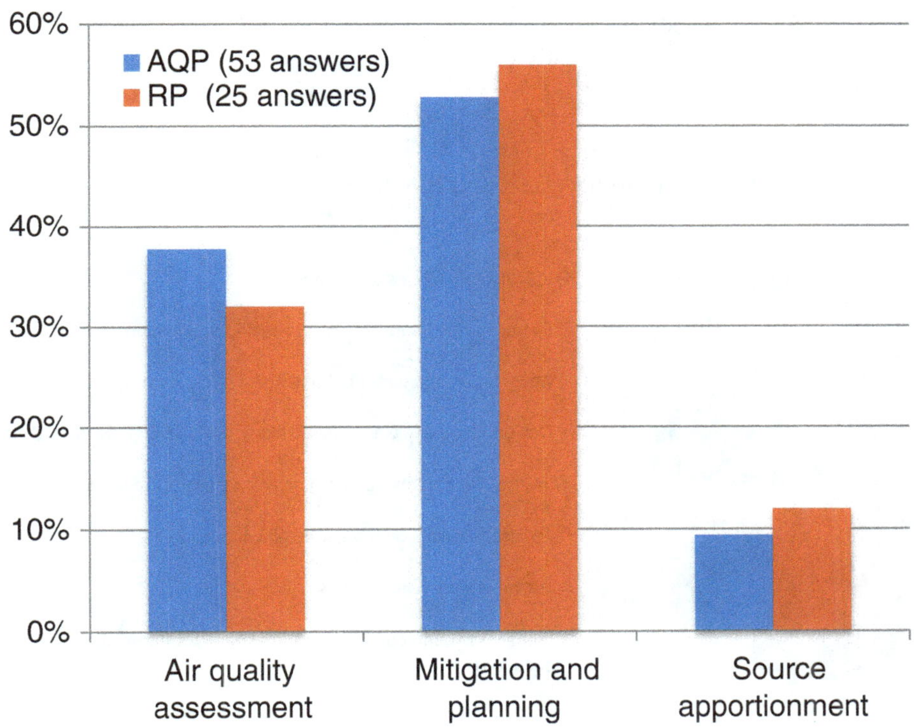

Fig. 3.12 Modelling purpose of AQPs (*blue*) and RPs (*red*)

health effects. Moreover source apportionment analysis within the framework of IA studies is applied to comply with the obligations deriving from the AQD, to design air quality plans or action plans, to identify the causes of exceedances, and to identify the transboundary pollution contribution from other countries (Fig. 3.14).

Receptor models and dispersion models (Lagrangian models, Eulerian models and Gaussian models) are used for the identification of sources. Objective

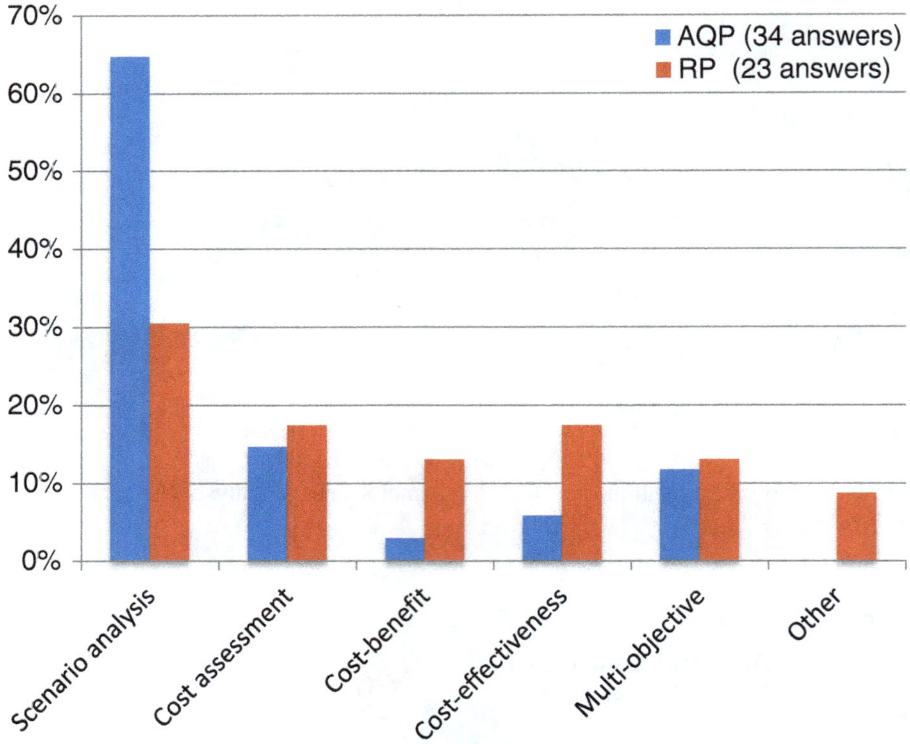

Fig. 3.13 IA methodologies in AQPs (*blue*) and RPs (*red*)

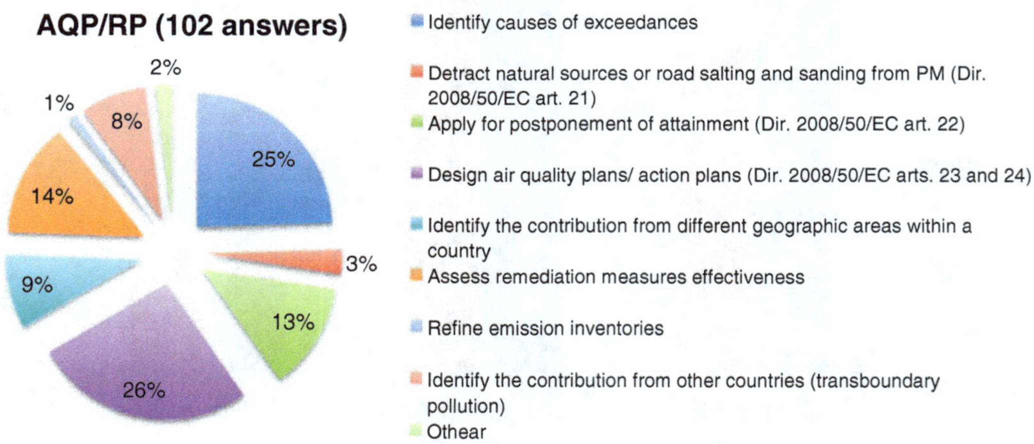

Fig. 3.14 Source apportionment purposes

estimation and inverse models are used marginally for this task (Fig. 3.15). It is worth to mention that one third of the answers report the combined use of more than one methodology.

The most frequent activity sectors/source categories identified in the studies are combustion in the energy sector and road transport (more than 70 % of the studies), followed by combustion in industry, non-industrial combustion and agriculture. Interestingly, many of the studies (40 %) focus only on one single activity

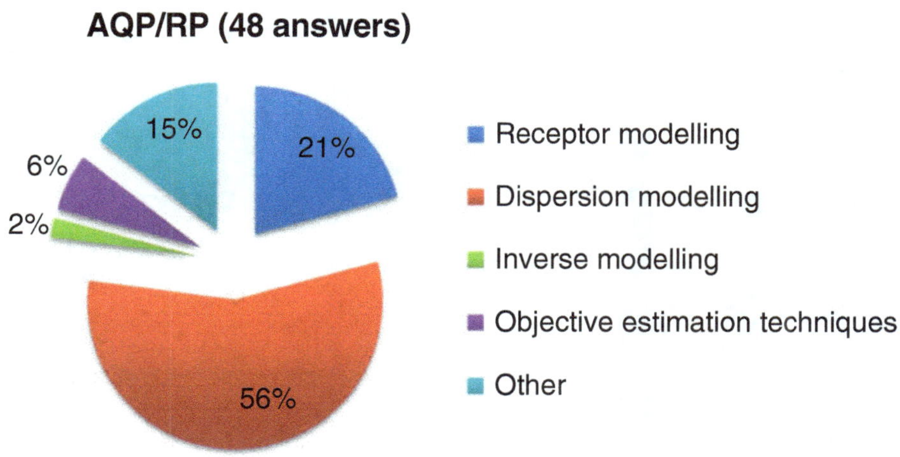

Fig. 3.15 Methodologies used for source apportionment

sector/source category. The frequency of such categories reflects the most commonly encountered pollution sources. Nevertheless, this is also influenced by the availability of source characterization studies and the existence of mandatory emission registers.

The most important pollutants considered in source apportionment studies are PM10 (84 %) and nitrogen dioxide (63 %) followed by two pollutants associated to them: PM2.5 (63 %) and nitrogen oxides (28 %), respectively. All the other pollutants are treated in less than 10 % of the studies.

The great majority of the studies focus on the city level (35 %) while local (lower than city) and regional scales represent a 32 and 22 % respectively. The country scale is marginally assessed (7 %).

The types of input data strongly depend on the adopted methodology. Monitoring networks and emission inventories are the most frequent sources of information (20 % each). Meteorological fields are input in 36 % of the answers while dedicated field campaigns represent the 16 %.

In the optimization approach, the emission reduction measures are selected by an optimization algorithm assessing their impact on air quality, health exposure, and implementation costs. Such optimization algorithms requires thousands of air quality assessments; in these cases, AQ systems cannot directly be used because of the computing time demand, so they provide tens to hundreds simulations processed to identify 'simple' emissions-AQ links (source/receptor relationships).

IAM approaches based on cost-benefit, cost-effectiveness or on multi-objective (i.e. optimization) approaches are used more often in research projects (61 %) than in AQPs (35 %). One explanation for this low proportion in the AQPs might be the fact that optimization approaches generally require extensive work to derive relationships to link emissions to air quality (source/receptor relationships) and to collect data related to emission reduction measures and costs and to externalities. Indeed these approaches cannot embed full 3D deterministic multi-phase modelling systems because of their prohibitive computational requirements.

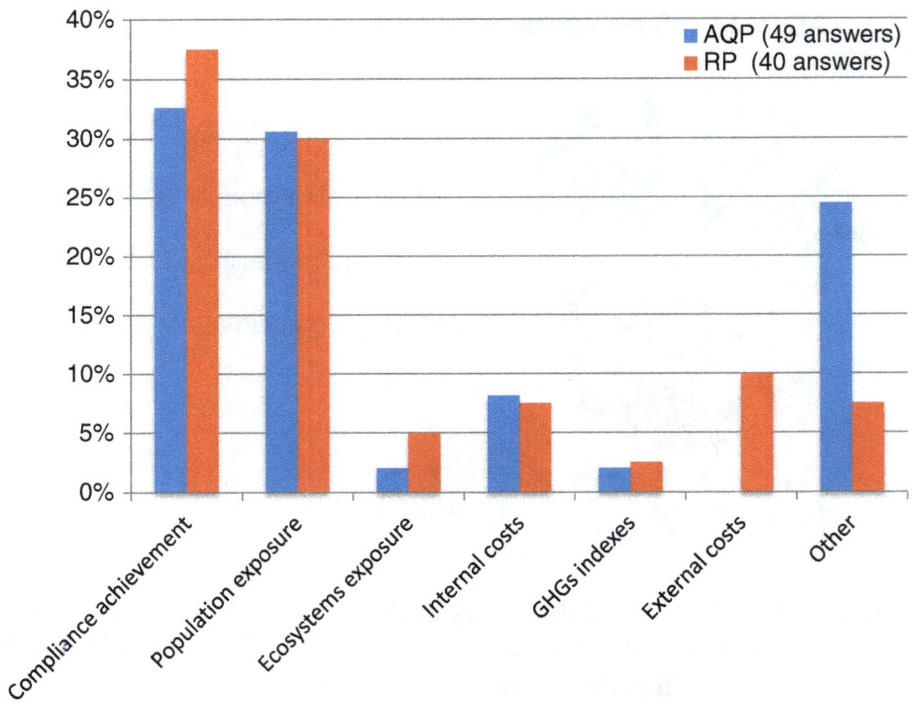

Fig. 3.16 Main indicators on which IAM tools focus

It is also interesting to assess which priorities were identified when designing air quality plans and running research activities. The reported priorities are focused on compliance achievement and population exposure followed by emission reduction costs (internal costs) and costs mainly related to the negative impact of air pollution on human health (external costs) (Fig. 3.16).

3.2.5 Scale and Resolution Issues

The synergies among national, regional and local approaches, including emission abatement policies, were analysed for the following aspects:

1. Contribution to decision level: 37 studies support the decision at the regional scale, 11 at the national scale and 31 at the local scale.
2. Emission sectors addressed with the AQ mitigation measures: Fig. 3.17 highlights the significance of SNAP 7 (Road traffic) and SNAP 2 (Non-industrial combustion) and the low involvement of SNAP 10 (agriculture) in defining policies. The traffic related emissions (SNAP 7, 94 %) were the focus of most AQPs with less prominent roles for non-industrial combustion (SNAP 2, 68 %). This is of course related to the pollutants targeted: most plans target nitrogen oxides for which traffic and combustion in general is the main source. For the

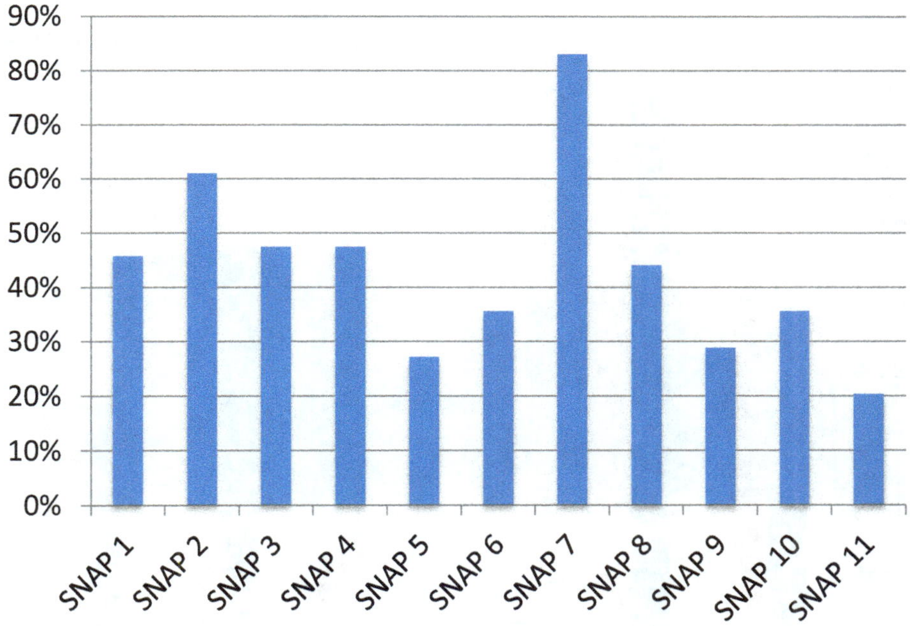

Fig. 3.17 Analysis of SNAP (SNAP1-combustion in energy and transformation industries; SNAP2-non-industrial combustion plants; SNAP3-combustion in manufacturing industry; SNAP4-production processes; SNAP5-extraction and distribution of fossil fuels and geothermal energy; SNAP6-solvent and other product use; SNAP7-road transport; SNAP8-other mobile sources and machinery; SNAP9-waste treatment and disposal; SNAP10-agriculture; SNAP11-other sources and sinks) sectors being addressed by air pollution measures

RPs, the attention to the different sectors is more equilibrated albeit also in this case SNAP 7 remains the most important sector.

3. Type of emission reduction measures: the number of non-technical and technical measures considered is very similar (41 and 39 % respectively).

3.2.6 Sensitivity and Uncertainty

Understanding the factors that contribute to the uncertainty in IA studies is quite complex. Out of the APPRAISAL database, 28 studies included responses to the topic on "uncertainty and robustness". The responses reported the current practise in quality control procedures when applying IAM for air quality related studies and AQPs. Out of these 28 responses, 14 were regarding to AQPs (41 % of the total AQPs) while 11 were RPs (61 % of RPs) and 3 represented other purposes.

In particular, the majority of model users rely on the operational evaluation technique (comparison with measurements) to assess the quality of the model results both in AQPs and RPs (Fig. 3.18). The other evaluation methods were also

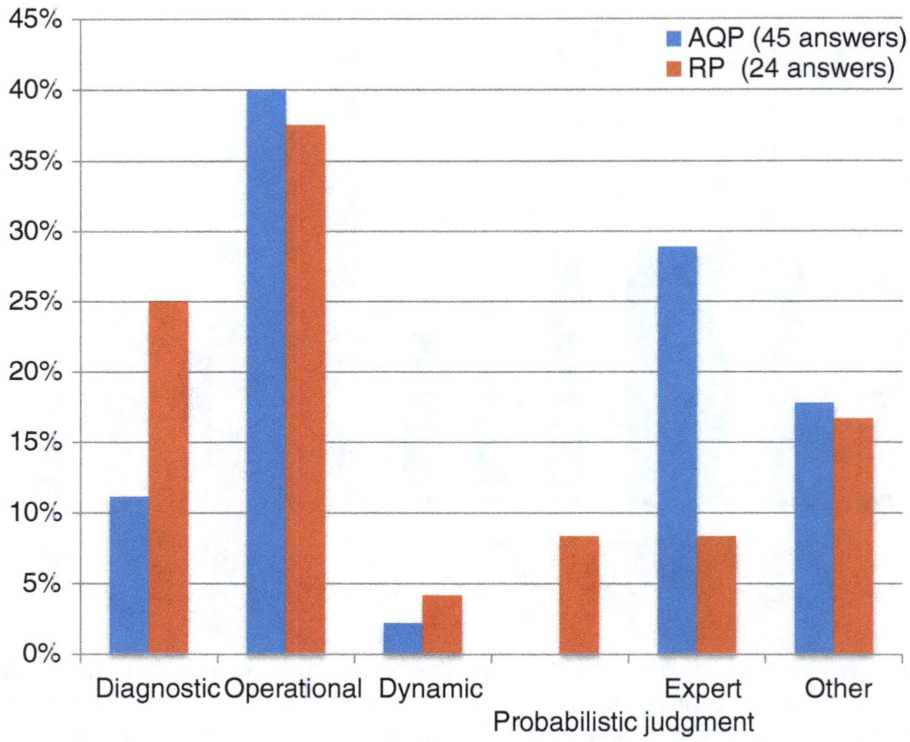

Fig. 3.18 Overview of evaluation methodologies used for the assessment of AQPs and RPs

represented in the returned questionnaires, although not so commonly applied. In the case of RPs, the percentage of responses indicating the use of a probabilistic or diagnostic method increases, whereas the number relying on expert judgement is relatively low. It can be therefore concluded, that a more comprehensive model evaluation process is performed in European member states in the frame of RPs than for AQP, with the operational evaluation dominating but complemented by other techniques. This can be attributed to the fact that these additional evaluation techniques require intensive personnel, infrastructure and time resources.

AQ modelling is the IAM component for which uncertainty analysis is most commonly considered in the questionnaire responses, both in the case of AQPs as well as for RPs (Fig. 3.19).

Nine of the responses reported that uncertainty estimation was performed for AQ modelling, one for source apportionment and 3 for health impact assessment, while uncertainty quantification for the IA system as a whole was represented only in 2 of the responses.

Global uncertainty analysis methods (e.g. Monte Carlo analysis) have been used in more studies compared to local uncertainty analysis methods more significantly, in RPs (Fig. 3.20). In some of the questionnaires, no answer was provided for the methodology used (local or global), particularly in the case of AQPs.

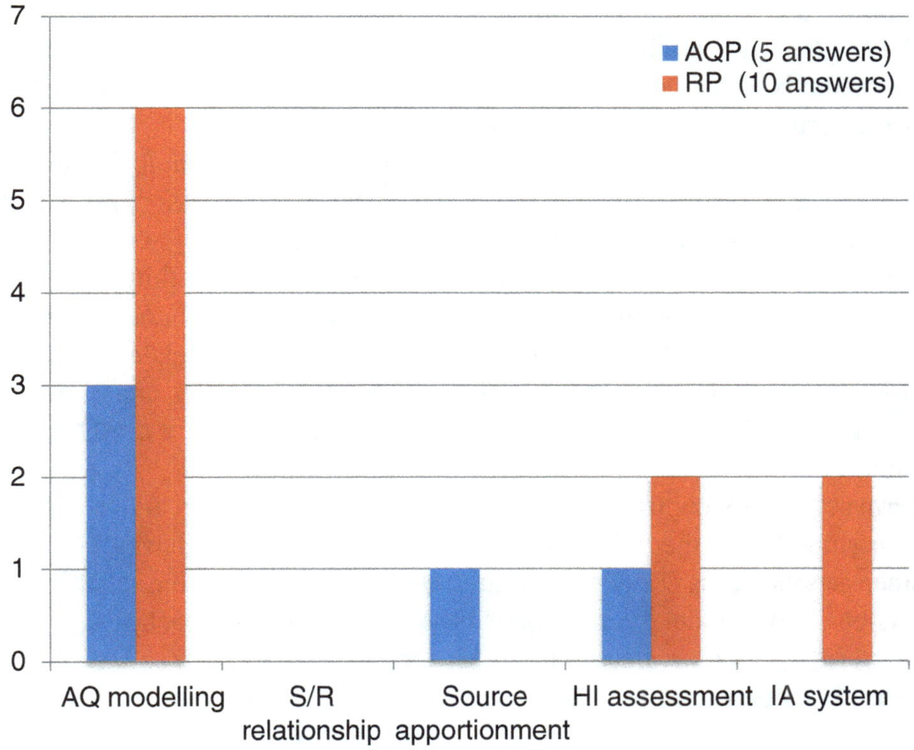

Fig. 3.19 Uncertainty estimation in different IAM components in AQPs and RPs

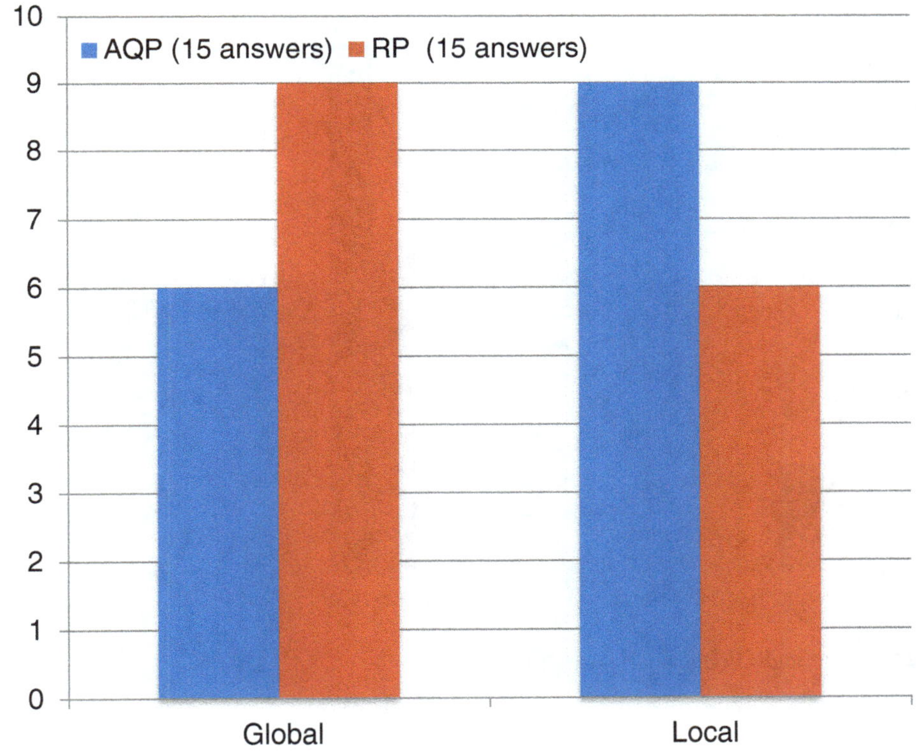

Fig. 3.20 Uncertainty analysis approaches in AQPs and RPs

Variance-based uncertainty estimation methods are the most commonly used among the global uncertainty assessment approaches. However, local uncertainty analysis methods (sensitivity methods, OaT) are also significantly represented in the responses, particularly in the case of RPs (Fig. 3.21).

The following Fig. 3.22 provides information on the AQ modelling elements for which uncertainty estimation was specifically carried out. As expected, model formulation was not one of the priority aspects examined in the case of AQPs; it was however considered in a significant number of RPs. Within AQPs, uncertainties were mostly analysed for meteorology, emissions and boundary conditions. Regarding RPs, it is interesting to note that uncertainties related to boundary conditions received less attention. For both AQPs and RPs, emissions related uncertainties are identified to significantly contribute to the total AQ modelling uncertainties.

In terms of quality control of model results for planning applications, most of the studies assumed that the AQ model is adequate when it behaves correctly for assessment applications (82 %) while in the 18 % of the cases the reliability of the model is based on model intercomparison and ensemble approaches.

It is interesting to note, that no reference technique is adopted so far to check the quality of the models used to quantify the impact of emission reduction scenarios in AQPs.

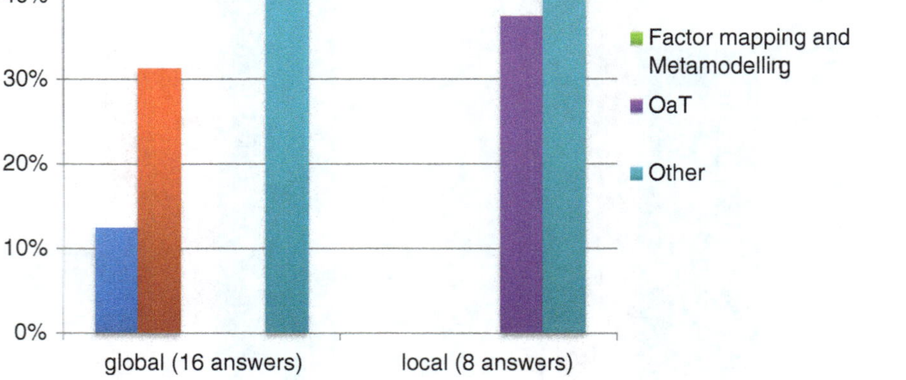

Fig. 3.21 Local and Global analysis methods

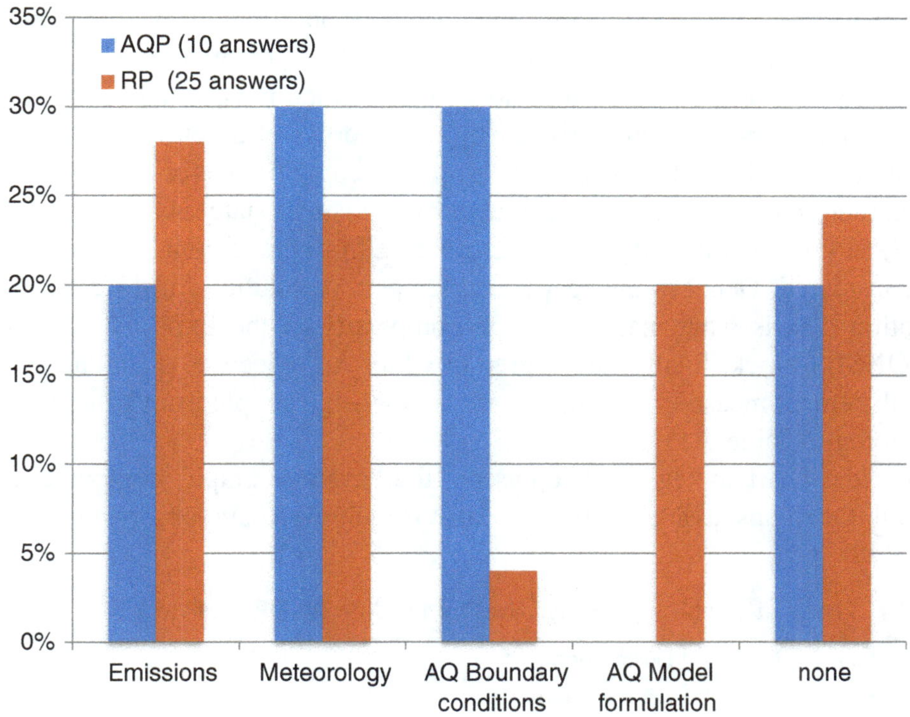

Fig. 3.22 Uncertainty estimation of different components of AQ modelling

3.3 Classifying the AQ Plans in Europe

The questionnaire responses have been classified trying to evaluate the level of detail at which each block of the DPSIR scheme has been treated. Though this classification is qualitative and partially subjective, it may serve a double purpose: within each plan, it highlights were more work has been invested and where, on the contrary, less attention was given; in comparison with other plans, it may indicate how a certain aspect has been dealt with in similar cases.

It must be noted that dealing with an aspect with a higher level of detail does not necessarily mean that the plan is more accurate or efficient in that field. Though the two things are hopefully correlated, there may be cases in which a more detailed approach was not supported by corresponding data or was not balanced with the corresponding costs or benefits.

The analysis of individual AQPs has been summarized using radar charts. This chart graphically represents the level of detail for each of the DPSIR blocks based on the answers to the questionnaire. For each of the five blocks, five levels of detail have been defined: Level 0—impossible to evaluate based on input from questionnaire (the topic is not even mentioned); Level 1—the block is considered in the AQP, but not investigated; Level 2—low level of detail in the implementation; Level 3—medium level of detail; and Level 4—high level of detail.

For the Driver block the complexity depends on whether the different levels (national, regional and local) are included as well as potential synergies between these different levels. For Pressure blocks the distinction is based on whether the activities and emissions were derived using a top down or a bottom-up approach or a combination of these two. The level of complexity for the state block (concentration/deposition) is determined by how the state is derived (using a model?) and whether the different scales ranging from the regional to the local scale were considered. Detail in the spatial and temporal resolution for the exposure and population data is what matters for the complexity of the Impact block. For the RESPONSES block, finally, the degree to which an objective, quantitative choice of the abatement measure(s) is made will distinguish a simple from a more complex methodology (Table 3.1).

The radar chart in Fig. 3.23 represents the "average graph" computed considering all the plans available in the database. Some main observations can be

Table 3.1 Levels of complexity distinguished for the different DPSIR blocks

DPSIR block	Level	Description
DRIVERS	1	not implemented
	2	top-down approach, using coarse spatial and temporal allocation schemes
	3	bottom-up approach with generic (i.e. national/aggregated) assumptions
	4	bottom-up approach with specific (i.e. local/detailed) assumptions
PRESSURES	1	not implemented
	2	emissions estimated for rough sectors on a coarse grid using a top-down methodology
	3	combination of bottom-up and top-down methodology
	4	emissions calculated with the finest resolution in space and time available (fine grid), using a bottom-up method and the highest level of detail in the SNAP sectors
STATE	1	not implemented
	2	measurements and geo-statistic interpolation are used
	3	one single deterministic model is used
	4	a downscaling nested models chain is used
IMPACT	1	not implemented
	2	coarse description of exposure provided either by measurement or modelling of AQ (e.g. average mean annual exposure for a city), simple population description
	3	similar to level 1, but with spatial detail in the STATE description
	4	detailed temporal and spatial resolution for exposure and population data
RESPONSE	1	not implemented
	2	expert judgment and scenario analysis
	3	source apportionment and scenario analysis
	4	Optimization

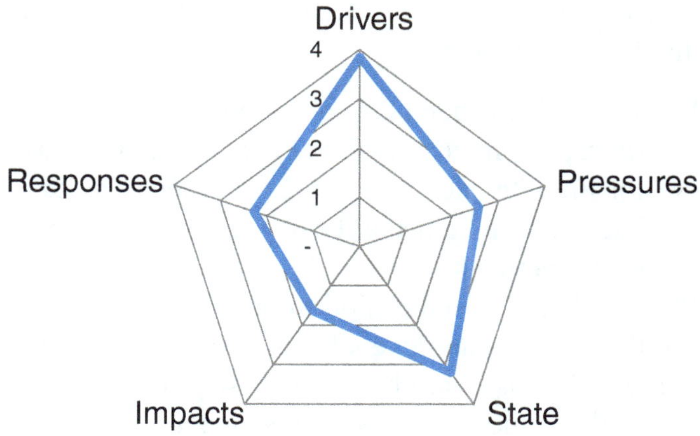

Fig. 3.23 "Average graph" computed considering all air quality plans available in the APPRAISAL database

derived. Most effort was put into quantifying the drivers and the state (concentration) in all the studies that were considered. The degree of detail used to evaluate emissions (PRESSURES) or to determine the consequent actions (RESPONSE) has been generally lower. Only rarely, actual plans and studies try to reach a quantification of the impacts on human health and ecosystems.

Following this approach, same examples of AQP classification are detailed in the next sections.

3.3.1 AQP for Athens

Description of the AQP

Athens AQP was developed as part of a wider effort of the Greek Ministry of Environment, Physical Planning and Public Works to comply to the EU legislation 1996/62/EC regarding ambient air quality levels. In this framework, the Ministry has funded the preparation of development plans for the abatement of air pollution in urban areas in Greece. For the urban area of Athens, the plan was jointly undertaken by two consulting companies, namely ENVECO S.A. and EPEM, with the official title: "Development of an Operational Plan for the Abatement of Atmospheric Pollution in the City of Athens".

The city of Athens is located in a basin of approximately 450 km^2. It is surrounded on three sides by fairly high mountains (Mt. Parnis, Mt. Pendeli, Mt. Hymettus and Mt. Aegaleon), while to the SW it is open to the sea. Industrial activities take place both in the Athens basin and in the neighboring Thriasion plain. The Athens basin is characterized by a high concentration of population (about 40 % of the Greek population), accumulation of industry (about 50 % of the Greek industrial activities) and high motorization (about 50 % of the registered Greek cars). Anthropogenic emissions in conjunction with unfavorable topographical and meteorological conditions are responsible for the high air pollution levels in the area.

The program for the abatement of air pollution in the urban area of Athens was divided into three phases:

Phase 1: This included the collection of emission data from all contributing sources (transport, industry, central heating) and the application of a dispersion model for the reference year 2002, in order to assess the spatial distribution of pollutants, complementarily to the measured concentration data from the monitoring network.

Phase 2: It included the application of an air quality dispersion model for predicting the air pollutant levels for the years 2005, 2008 and 2010.

Phase 3: In this final phase, a Decision Making System was developed in order to evaluate the efficiency of abatement measures in terms of compliance with the EU Directive.

Drivers/Pressures

The main drivers identified included industry, central heating and transport. However, in terms of PM10, an additional source apportionment study was performed which included sources particularly linked to PM10 emissions, such as long-range transport and resuspension.

Within the development of the AQP, the Greek Ministry of Environment funded the compilation of an emission inventory which was compiled for the Greater Athens Area, for the reference year 2002, taking into account emissions from:

1. Stationary air pollution sources like, industry, domestic heating and oil stations,
2. Mobile sources, such as, road traffic and emissions from ship, airplane and train lines.

Pollutants included were CO, NO_2, NOx, O_3, SO_2, Benzene, PM10 and Pb, for most of which EU legislation sets up specific air quality limit values that had to be met within 2005 and 2010. Regarding stationary air pollution sources, an on-site measurement campaign was undertaken including 1000 industrial units from 48 industrial sectors. An emission factor database adapted for Greece was also prepared. Concerning the emission inventory for road traffic emissions, the CORINAIR methodology (EEA 2013) and the COPERT software (COPERT4 2007) were applied. A detailed bottom-up emission inventory was the result of this effort.

Emission rates for pollutants from transport and industry were derived from the National Emission Inventory (Ministry of Environment), while biogenic emissions were based on existing published results. The emission rates for tire wear, brake wear and road abrasion were calculated based on the CEPMEIP database (http://www.air.sk/tno/cepmeip/), while the construction activity was approached from satellite images and traffic resuspension emissions from literature data.

State

In this AQP, both air quality assessment as well as a source apportionment methodology for PM10 were applied.

Regarding the urban air quality assessment, it can be concluded that this was addressed at an advanced complexity level. The Eulerian OFIS urban scale dispersion model (Moussiopoulos and Sahm 2000) was used for the spatial assessment of pollutant levels in the study area and for the development of maps allowing the identification of heavily polluted areas within the study domain. OFIS simulates concentration changes due to the advection of species and chemical reactions in each cell of the computational domain. In order to account for the contribution from local emission sources, the OSPM combined plume and box model (Berkowicz et al. 2008) was used for simulations of air pollution from traffic in urban streets.

The influence of meteorological patterns on PM10 concentrations was analyzed, particularly in regard to long-range PM10 transport from other areas (e.g. the Saharan desert). The contribution of natural sources was assessed using a combined methodology of satellite images, LIDAR measurements, measurements from the national monitoring network and modelling results using the SKIRON/Eta transport and deposition model (Kallos et al. 1997)

Concentrations of pollutants were assessed using a chain of models adapted to different scales from the regional to the local scale. The Eulerian model OFIS takes into account regional background pollutant levels to evaluate the transfer of pollutants towards and away from the urban area. Furthermore, all main chemical transformation mechanisms are represented in the OFIS model, which is a pre-requisite for studying reactive pollutants such as ozone and particles. The OSPM street scale model accounts for increased concentrations at the local (hot-spot) scale due to local emissions. Both models have an appropriate spatial and temporal resolution to realistically describe pollutant dispersion at the scales of interest. Furthermore, both a sensitivity analysis in terms of emissions was conducted (emission reduction scenarios and sensitivity to natural background contributions) as well as an operational model validation against measurement data from the monitoring network. In conclusion: an advanced (Level 3) complexity level was used for concentration assessment.

Impact

The impact of the assessed pollutant concentration levels on health was not specifically addressed in the development of this AQP. This parameter was only indirectly considered, on the basis of exceedances of limit values for the protection of human health, according to the EU Directive.

Response

The simulations were performed for the urban scale as well as for the street scale model for several emission scenarios, for the years 2005, 2008 and 2010, in order to examine compliance with standards.

The results indicated that natural emission sources play a very important role in the calculation of PM concentrations and that their contribution leads to significant increase in the number of current and future exceedances. This could suggest that stricter policies regarding the anthropogenic part of PM emission need to be applied.

A source apportionment study was conducted for PM10. The spatial and temporal distribution of PM10 in the Greater Athens Area was assessed with the use of the Eulerian photochemical model REMSAD (S.A.I. 1998) and sensitivity simulations were performed with the same modelling tool to identify and quantify source contribution.

An interesting point in the AQP for Athens was that different emission reduction scenarios were evaluated both for the urban scale (using the OFIS model) as well as for particular hot spots due to local traffic emissions (using the OSPM model). In this way it was shown that a further emission reduction is required in order to comply with standards at the local scale (i.e. to reduce number of exceedances), on top of the emission reduction that is necessary to comply with annual limit values.

An optimization procedure was not performed. A thorough Multiple Criteria Analysis using the ELECTRE III method (Roy 1968) was applied in order to identify the most suitable set of abatement measures. Parameters such as the public cost, public acceptance and socio-economic impacts were considered.

The overall plan may thus be represented by the chart in Fig. 3.24.

3.3.2 *AQP for Emilia Romagna*

Description of the AQP

This study was concerned with the Po Valley area and in particular with the Emilia-Romagna region. The aim of the study was mainly to assess the benefits of different sets of measures to improve air quality.

The Emilia-Romagna region is located in the south-western part of the Po Valley basin, a densely populated and heavily industrialized area, where meteorological conditions, due to the low wind intensity, cause the stagnation of the air masses, associated with peak pollution episodes of PM during winter time and high levels of ozone during the summer time. The daily Limit Value (LV) for PM10 was exceeded every year since the enforcement of the EU directive (2008) with a slow decreasing trend of the PM10 annual mean during 2001–2012. The NO_2 annual limit value shows some exceedances mainly at the traffic stations and a decreasing

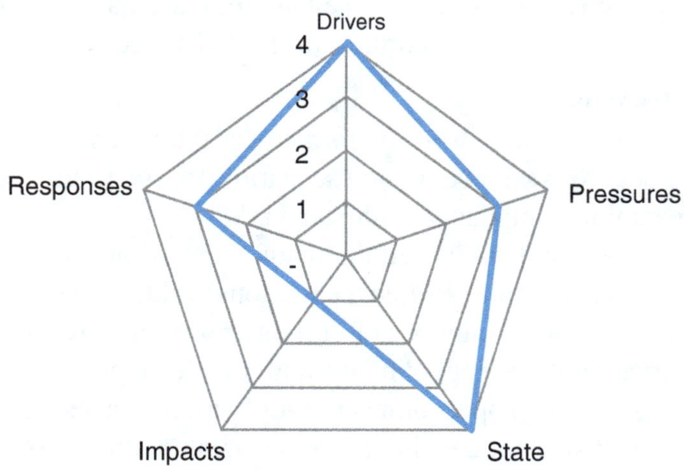

Fig. 3.24 Radar chart for the AQP of Athens (levels: 4 = high; 3 = medium; 2 = low; 1 = not considered; 0 = N/A)

trend. Ozone health and vegetation protection limit values are systematically exceeded in all the stations with a stationary trend during 2001–2012. The data show also that the annual LV for PM2.5 (obligation from 2015) can be exceeded with adverse meteorological conditions.

Drivers/Pressures

Sources of PM and ozone precursors, such as NOx and VOCs, are mainly related to road transport and combustion. Almost 60–65 % of particulate matter is of secondary origin and a large part of particulate matter and ozone pollution is due to regional background that is influenced by the transport of pollutants from the neighboring regions of the Po Valley basin. NO_2 exceedances are mainly due to local pollution, nevertheless the background concentration of NOx plays an important role in the production of the secondary aerosols. Ammonia (which is mainly emitted by agriculture) is an important precursor of PM in the Po Valley. Diesel trucks are responsible for a large part of NOx emissions. Emissions from wood burning and motor vehicles (exhaust and non-exhaust) are the main sources of PM10.

The emission scenarios and the resulting air pollution simulations have been produced on a domain grid covering the Emilia-Romagna region and the surrounding areas, which influence the regional air pollution. The regional inventory of atmospheric emissions has been undertaken by regional environmental agency (ARPA-ER) on behalf of the Emilia-Romagna Region, with reference to the year 2010 using INEMAR (INventario EMissioni in ARia—Air Emission Inventory, http://www.inemar.eu/xwiki/bin/view/Inemar/WebHome): a data collection and processing system developed to guide the development of a regional bottom-up atmospheric emission inventory for different activities (heating, road transport, agriculture, industry, etc.). The gridded emissions and proxy variables were prepared using the tool eFESTo, which is part of the NINFA Regional Air Quality Modeling System (Stortini et al. 2007). This input allows the RIAT+ tool (Carnevale et al. 2012) to produce a spatial and seasonal disaggregation of the emissions inside the region.

The regional emission inventory details emissions by macro sector-sector-activity and fuel (inside the Region); the point source emissions also have stack details.

State

To determine NO_2, PM and O_3 related AQIs a nested chain of Eulerian models was used. Air pollution concentrations have been simulated for the year 2010 using NINFA, which includes CHIMERE (version 2008c), a Eulerian chemical transport model. The range of scale was regional and urban; the spatial resolution was 5 km by 5 km, with 40 vertical levels; the output consists of hourly concentrations. The meteorological model used is COSMO17 (http://www.cosmo-model.org), with a prognostic approach. The background contribution was determined as hourly concentrations using the Prev'air model (http://www.prevair.org/en/modele.php). The concentrations due to the local traffic/industry emissions were then further refined to street level.

Emission data (for NOx, VOC, NH_3, PM10, PM2.5, SO_2) and AQI computed values (mean PM10, mean PM2.5, AOT40, SOMO35, mean NO_2, mean MAX8H O_3) have been then used to train the Artificial Neural Networks (ANNs), which describe the relationship between emissions of the precursors and the AQI for each temporal period (year, winter and summer). The results confirmed that the neural network surrogate model is capable of reproducing the non-linear relationship between emissions and precursors.

To train the ANNs, 12 emission scenarios on the Emilia-Romagna domain were designed and used.

Impact

For the health impact assessment, the high-resolution concentration maps were combined with a detailed population map. The approach used was retrospective. The health impact relationship used dealt with the reference values associated to the relative risks, without thresholds. Population data used for the health impact functions originated from a cohort study. The air pollutants used in the estimation were: PM2.5, Arsenic, Cadmium, Nickel and other. The exposure indicators were calculated based on interpolated monitored data and modeled values. For population, the same spatial and temporal resolution of concentration were used. The indicator used was the morbidity (e.g. pneumonia cases, cardiovascular and respiratory diseases).

Response

In this preliminary phase of the Regional AQP, the RIAT+ tool has been used to assess measures and costs to improve air quality. Both technological and efficiency measure are taken into account in the optimization process. Analyzing the yearly average PM10 concentration on the whole Emilia-Romagna, a Pareto curve was obtained, the points of which represents different optimal combinations of reduction measures. The analysis of the Pareto curve shows that a significant reduction of NH_3 should be reached acting on agriculture macro sector, while NOx reduction should be obtained through transport and other mobile sources macro-sectors. Actions on residential heating should be promoted to reduce a large part of primary PM10 component.

RIAT+ gave also a detailed list of measures to obtain these reductions. The combination of different runs with single or multi-pollutant optimization objectives leads to the following list of priority measures to be implemented:

- Energy efficiency measures in the residential sector including improved fireplaces;
- High efficiency oil and gas industrial boilers and furnaces in manufacturing industry;
- Significant replacement of old heavy and light duty diesel vehicles with newer Euro5 and Euro6 compliant), as well as an increase of the limited traffic zones and cycling paths;
- Replacement of oldest construction and agriculture vehicles.

The overall plan may thus be represented by the chart in Fig. 3.25.

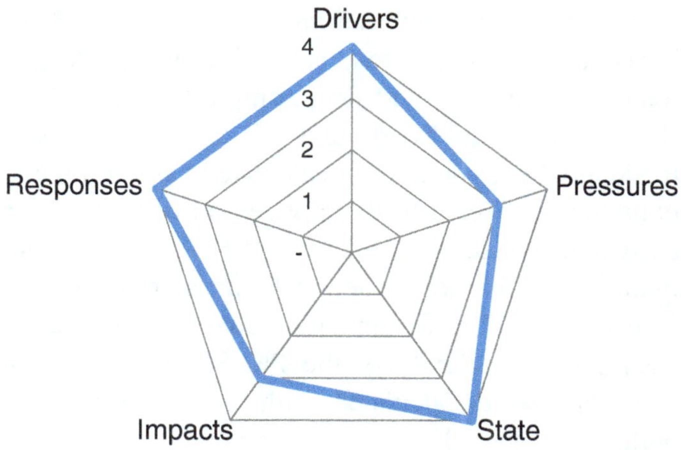

Fig. 3.25 Radar chart for Emilia-Romagna preliminary AQP (levels: 4 = high; 3 = medium; 2 = low; 1 = not considered; 0 = N/A)

3.3.3 AQP for the Warsaw Agglomeration

Description of the AQP

Warsaw has about 1.7 million inhabitants and is the largest and one of the most congested cities in Poland. This is mainly due to the lack of a real bypass road, so most of the traffic is routed through city streets, which are quite narrow in many areas. The Warsaw metro is one of the newest subway systems in Europe, however it has only one line so far. Building activities for the second line—which is being currently realized—constitutes an additional disruption in city traffic. In general, bicycle routes are scarce, being well organized only in a few districts. As a result, according to the latest assessment (Deloitte 2014) each Warsaw's dweller loses on average a month of salary a year, due to time spent in traffic congestion.

The first Air Quality Plan for Warsaw was issued due to the exceedances of PM10 and NO_2 limit values in 2004. The road transport sector (SNAP07) has the biggest share in all pollutants concentrations, but there are a few districts with a significant share of residential heating. In general, the contribution of transport emissions to PM concentrations is constantly growing. Beyond the exceedance zones, the pollutants inflow from outside of the agglomeration has an important share, at times being the prevailing one.

This AQP study was performed for the years 2004–2007. Furthermore, plans concerning B(a)P (2007) and PM2.5 (2010) were also established. Warsaw agglomeration zone is considered as a hot spot with problems in terms of exceedances of the NO_2 and PM guidelines of the EC Directive. A new AQP is currently being implemented (up to the end of 2016).

Drivers/Pressures

The Air Quality Plan (AQP) for Warsaw takes into account national, regional and local strategies and applies bottom-up approach, therefore the complexity of the DRIVERS block is high (level 4). The main local activities are: road transport, residential heating, energy production and industry.

The emission database was generated by EKOMETRIA Agency. For traffic, hourly emissions for a road network were calculated as a function of traffic volume, road characteristics and fleet composition, based on the data from the Warsaw's Boards of Urban Roads and of Public Transport (250 m × 250 m resolution). Residential emissions were calculated based on the local information on residential units not connected to the city central heating system, their furnace type and fuel used (coal, coke, gas, oil, wood) (250 m × 250 m resolution, as well). For the industrial emissions a detailed emission inventory (compiled by the Regional Inspectorate of Environmental Protection in Warsaw) with stack level data was used. The complexity of the PRESSURES block is thus also high (level 4) as emissions were calculated with a fine resolution in space and time, using a bottom-up method.

State

To determine the NO_2 and PM10 concentrations a chain of models was used. The concentrations for the study area (covering the agglomeration and its 30 km diameter surroundings) were calculated with a CALPUFF (http://www.src.com) Gaussian puff model setup (discrete receptors were used) with decreasing resolution from 1 km (for city surroundings) to a very high 250 m resolution (for the agglomeration itself). Regional (Voivodeship) background concentrations were calculated at a resolution of 7 km using the CAMx Eulerian chemical transport model (Environ 2006) and included monthly varying boundary conditions also for aerosols derived from the EMEP Unified model (50 km resolution, monthly averages).

Operational model evaluation was carried out with the set of statistical metrics proposed by Juda-Rezler et al. (2012).

The features of CALPUFF model also allowed to compute the contribution of different source categories to the air pollution in the study area (source-apportionment).

In summary, the level of complexity of the STATE block can be considered high (level 4).

Impact

In the AQP for Warsaw, the human health effects were not directly considered, and indirectly measured, as determined by the exceedances of limit values for the protection of human health, according to the EU Directive. The analysis was based on yearly average concentrations for NO_2 and both yearly and daily averages for PM10 concentrations. Thus, the IMPACT assessment block level is 1.

Response

In this study a preliminary list of economically and/or socially and politically feasible measures was drafted and subsequently extended and screened based on expert opinion and previous experience with respect to the effectiveness of the individual measures. Besides the measures, also a map of hot spots was provided for which the measures should be applied. The finally proposed measures were split into two groups:

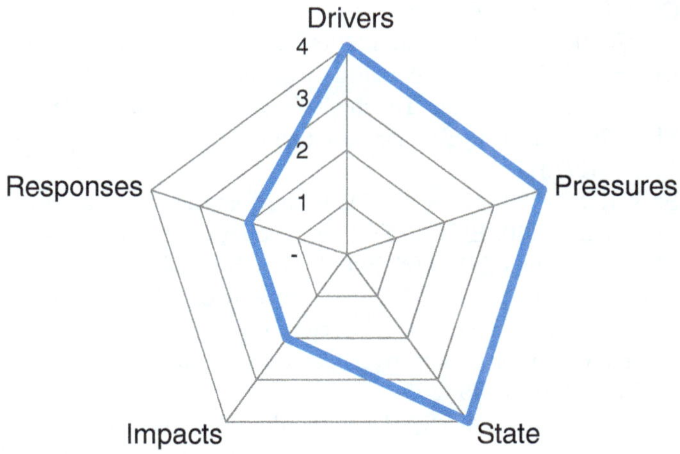

Fig. 3.26 Radar chart for the AQP for Warsaw Agglomeration (levels: 4 = high; 3 = medium; 2 = low; 1 = not considered; 0 = N/A)

1. Measures to be implemented to the residential emission:
2. Connection of individually heated houses to the municipal heating network: This measure is proposed for 4 districts, covering approximately 1 % of the agglomeration area, with approximately 13,000 inhabitants.
3. Measures to be implemented to the road transport emission: Improvement of public transport network by building of 2 ring roads: City Centre Ring Road & City Ring Road (up to 2020) and establishment of a low emission zone in the City Centre.

Implementation of the first measure alone will reduce total PM10 emission in the zone by as much as 21 %, while implementation of the second will reduce total PM10 and NO_2 emissions in the zone by 30 and 53 %, respectively.

For each of proposed measures differences in concentration were calculated (scenario analyses).

The study did not use either source apportionment or an optimization procedure to derive the set of abatement measures Given that the RESPONSES block is based on expert judgment and scenario analyses, it complexity appears to be relatively low (level 2) and the overall AQP chart may be represented as in Fig. 3.26.

Acknowledgments This chapter is partly taken from APPRAISAL Deliverables D2.2, D2.3, D2.4, D2.5, D2.6, D2.7, D4.2 (downloadable from the project website http://www.appraisal-fp7.eu/site/documentation/deliverables.html).

References

Berkowicz R, Ketzel M, Solvang JS, Hvidberg M, Raaschou-Nielsen O (2008) Evaluation and application of OSPM for traffic pollution assessment for a large number of street locations. Environ Model Softw 23:296–303

Carnevale C, Finzi G, Pisoni E, Volta M, Guariso G, Gianfreda R, Maffeis G, Thunis P, White L, Triacchini G (2012) An integrated assessment tool to define effective air quality policies at regional scale. Environ Model Softw 38:306–315

COPERT4 (2007) Emission inventory guidebook, chapter road transport, activities 070100-070500, Version 6.0 of 23 August 2007. Aristotle University Thessaloniki, Greece

Deloitte (2014) Traffic congestion in the 7 biggest cities in Poland. Report of Deloitte Poland, Member of Deloitte Touche Tohmatsu Limited (In Polish)

EEA (2013) EMEP/EEA air pollutant emission inventory guidebook 2013. EEA Technical report No 12/2013

ENVIRON (2006) User's guide to the comprehensive air quality model with extensions (CAMx), Version 4.40. ENVIRON International Corporation, Novato, California

EU (2008) Directive 2008/50/EC on ambient air quality and cleaner air for Europe

Juda-Rezler K, Reizer M, Huszar P, Krüger BC, Zanis P, Syrakov D, Katragkou E, Trapp W, Melas D, Chervenkov H, Tegoulias I, Halenka T (2012) Modelling the effects of climate change on air quality over Central and Eastern Europe: concept, evaluation and projections. Clim Res 53:179–203

Kallos G, Nickovic S, Papadopoulos A, Jovic D, Kakaliagou O, Misirlis N, Boukas L, Mimikou N, Sakellaridis G, Papageorgiou J, Anadranistakis E, Manousakis M (1997) The regional weather forecasting system SKIRON: an overview. In: Proceedings of the international symposium on regional weather prediction on parallel computer environments, 15–17 Oct 1997, Athens, Greece, pp 109–122

Moussiopoulos N, Sahm P (2000) The OFIS model: an efficient tool for assessing ozone exposure and evaluating air pollution abatement strategies. Int J Environ Pollut 14:597–606

Parra MA, Santiago JL, Martín F, Martilli A, Santamaría JM (2010) A methodology to assess urban air quality during large time periods of winter using computational fluid dynamic models. Atmos Environ 44:2089–2097

Roy B (1968) Classement et choix en présence de points de vue multiples (la méthode ELECTRE). La Revue d'Informatique et de Recherche Opérationelle (RIRO) 8:57–75

S.A.I. (1998) User's guide to the regulatory modeling system for aerosols and deposition (REMSAD), SYSAPP98-96\42r2. Systems Applications International Inc, San Rafael

Stortini M, Deserti M, Bonafè G, Minguzzi E (2007) Long-term simulation and validation of ozone and aerosol in the Po Valley. In: Borrego C, Renner E (eds) Developments in environmental sciences. Elsevier, Amsterdam, vol 6, pp 768–770

Chapter 4
Strengths and Weaknesses of the Current EU Situation

C. Belis, N. Blond, C. Bouland, C. Carnevale, A. Clappier, J. Douros, E. Fragkou, G. Guariso, A.I. Miranda, Z. Nahorski, E. Pisoni, J.-L. Ponche, P. Thunis, P. Viaene and M. Volta

4.1 Introduction

As already noted, the 2008 European Air Quality Directive (AQD) (2008/50/EC) encourages the use of models in combination with monitoring in a range of applications. It also requires Member States (MS) to design appropriate air quality

C. Belis · G. Guariso
Politecnico di Milano, Milan, Italy

N. Blond
Centre National de la Recherche Scientifique (CNRS), Paris, France

C. Bouland
Université Libre de Bruxelles (ULB), Brussels, Belgium

C. Carnevale · M. Volta (✉)
Università degli Studi di Brescia, Brescia, Italy
e-mail: marialuisa.volta@unibs.it

A. Clappier ·
J.-L.Ponche
Université de Strasbourg, Strasbourg, France

J. Douros · E. Fragkou
Aristotle University of Thessaloniki, Thessaloniki, Greece

A.I. Miranda
Universidade de Aveiro, Aveiro, Portugal

Z. Nahorski
Systems Research Institute of the Polish Academy of Sciences (IBS-PAN),
Warsaw University of Technology, Warsaw, Poland

E. Pisoni · P. Thunis
European Commission, Joint Research Centre (JRC), Directorate for Energy,
Transport and Climate, Air and Climate Unit, Ispra, Italy

P. Viaene
Vlaamse Instelling voor Technologisch Onderzoek n.v. (VITO), Mol, Belgium

G. Guariso and M. Volta (eds.), *Air Quality Integrated Assessment*,
PoliMI SpringerBriefs, DOI 10.1007/978-3-319-33349-6_4

plans for zones where air quality does not comply with the AQD limit values and to assess possible emission reduction measures to reduce concentration levels. These emissions reductions then need to be distributed in an optimal and cost effective way through the territory. Obligations resulting from other EU directives (e.g. the National Emission Ceiling Directive) and targeting more specific sectors of activity (e.g. transport, industry, energy, agriculture) must also be considered when designing and assessing local and regional air quality plans (Syri et al. 2002; Coll et al. 2009). In order to cope with these various elements MS have in the last decade developed and applied a wide range of different modelling methods to assess the effects of local and regional emission abatement policy options on air quality and human health (e.g. Cuvelier et al. 2007; Thunis et al. 2007; De Ridder et al. 2008; Carnevale et al. 2011; Lefebvre et al. 2011; Borrego et al. 2012; Mediavilla-Sahagun and ApSimon 2013).

4.2 Available Tools

The following Table 4.1 summarizes the integrated assessment modelling tools most used in European countries. They can be classified in different ways according to the blocks of the DPSIR framework they investigate deeper, and are based on data collected from various public and specific sources.

At the EU level, the state-of-the-art regarding decision-making tools is GAINS (Greenhouse Gas and Air Pollution Interactions and Synergies), developed at the International Institute for Applied Systems Analysis, Laxenburg, Austria, by Amann et al. (2011). The GAINS model considers the co-benefits of simultaneous reduction of air pollution and greenhouse gas emissions. It has been widely used in international negotiations (as in the 2012 revision of the Gothenburg Protocol) and is currently applied to support the EU air policy review. Some national systems have been developed, starting from the GAINS methodology at EU level. Two well-known implementations are RAINS/GAINS-Italy (D'Elia et al. 2009) and RAINS/GAINS-Netherlands (Van Jaarsveld 2004). Another national level implementation is the FRES model (Karvosenoja et al. 2007), developed at the Finnish Environment Institute (SYKE) to assess, in a consistent framework, the emissions of air pollutants, their processes and dispersion in the atmosphere, effects on the environment and potential for their control and related costs. An additional important initiative at national level is the PAREST project, in which emission reference scenarios until 2020 were constructed for PM and for aerosol precursors, for Germany and Europe (Builtjes et al. 2010). The ROSE model (Juda-Rezler 2004) has been developed at Warsaw University of Technology (WUT) for Poland. ROSE is an effect-based IAM comprised of a suite of models: an Eulerian grid air pollution model, statistical models for assessing environment sensitivity to the Sulphur species and an optimization model with modern evolutionary computation techniques.

Table 4.1 Major IAM tools used in Europe

		Models/methodologies	Databases
Drivers and Pressures	Emission models	EMEP/EEA, INEMAR, COPERT-IV, TREM, HBEFA, IMMISem, HERMES, EC4MACS, TNO-MACC	– UNECE LRTAP Convention and NEC Directive inventories and projections – European Pollutant Release and Transfer Register (E-PRTR) – MACC/TNO emission inventory from the Copernicus Atmosphere services – EDGAR (Emissions Database for Global Atmospheric Research) – National/regional/urban emission inventories
State	Online and offline chemical transport models at regional scale	CHIMERE, CAMx, TCAM, MEMO/MARS-aero, WRF-Chem, CMAQ, EMEP, POLYPHEMUS	– AQ reporting database (before 2015: AirBase), EMEP, EBAS databases of observations – Copernicus archives for model results
	Air quality models at urban scale	ADMS, URBAIR, IMMIS, MISKAM, OFIS, VADIS	– AQ reporting database of observations
	Air quality and Climate Change Global scale models	WITCH, TM5-FASST, MOZART-4, MACC-IFS, NMMB/BSC	– GAW network – Satellite observations (available from the Copernicus services) – Copernicus archives for model results – ENER – IEA, EDGAR
	Regional Climate models	EURO-CORDEX Ensemble of Regional Climate Models	Earth System Grid Federation data nodes Inputs: ERA-Interim and CMIP5 Global Earth System Model future climate projections
	Source/receptor models	EMEP S/R matrices, ANNs	FAIRMODE DB
	Short term forecast	Deterministic, Data driven models	Urban databases

(continued)

Table 4.1 (continued)

		Models/methodologies	Databases
Impact	Monitoring network assessment	Optimal allocation algorithms Spatial representativeness methodology	Ground level observations (AQ reporting database, EMEP, EBAS) Satellite observations (Copernicus) Individual exposure databases
	Health assessment	BenMAP-CE, ARP, RIAT+, CENSE	– UN population forecasts – Concentration-response functions from the CAFE and HRAPIE studies
	Social acceptance	Discrete Choice Models (DCMs), Randomized Control Trials (RCT)	EUROBAROMETER local surveys, SEFIRA FP7 database
	Economic assessment	Labour Productivity Impact (LPI) analysis, RIAT+	Eurostat, NUTS (1,2,3), city economic, population data, European and international trade data, local business data
Responses	Source-apportionment	CAMX (OSAT/PSAT), CMAQ-ISAM PMF, CMB	FAIRMODE S/R fingerprints
	Source sensitivity	CAMX(DDM), HDDM-CMAQ	
	Scenario analysis	CHIMERE, TRANS TOOLS, RIAT+	
	Cost-effective analysis	RIAT+, WITCH integrated assessment model	GAINS, economic, population data
	Cost-benefit analysis	LPI analysis, cost curves, WITCH integrated assessment model	GAINS, economic, population data
	Multi-objective analysis	RIAT+	GAINS, economic, population data
	Multi-criteria analysis	ELECTRE, ELECTRE-III, AHP	

At urban/local scale a few integrated assessment models have been developed and applied (e.g. Vlachokostas et al. 2009; Zachary et al. 2011; Mediavilla-Sahagun and ApSimon 2013). In RIAT (Carnevale et al. 2012) the main goal is to compute the most efficient mix of local policies required to reduce secondary pollution exposure, in compliance with air quality regulations, while accounting for characteristics of the area under consideration. RIAT solves a multi-objective optimization, in which an air quality index is minimized constrained by a specific emission reduction implementation cost. It will be described in more details in the following chapter. The Luxembourg Energy Air Quality model (LEAQ) (Zachary et al. 2011) integrated assessment tool focuses on projected energy policy and related air quality at the urban and small-nation scale. The tool has been developed initially for the Grand Duchy of Luxembourg, but is flexible and could be adapted for any city with sufficient information concerning energy use and relevant air quality. The UKIAM model (Oxley et al. 2003) has been developed to explore attainment of UK emission ceilings, while meeting other environmental objectives, including urban air quality and human health, as well as natural ecosystems. Nested within the European scale ASAM model (Oxley and ApSimon 2007), UKIAM operates at high resolution, linked to the BRUTAL transport model for the UK road network to provide roadside concentrations, and to explore non-technical measures affecting traffic volumes and composition.

4.3 Areas for Future Research of DPSIR Blocks

This section identifies limitations of the current assessment methods and proposes key areas to be addressed by research and innovation. It is organized into several sub-sections, each corresponding to a specific building block of the DPSIR scheme.

4.3.1 Drivers (Activities)

Considerable weaknesses were identified for the DRIVERS block, for all activity sectors contributing to local scale emissions: not only for power plants, road traffic and residential combustion, but also agriculture, non-road traffic and machinery. An important future research line should be devoted to the integration of activity inventories at different scales. At the moment, inconsistencies exist between local/regional and EU level data collection methods and tools, and this prevents the implementation of a fully integrated approach connecting the various governance scales. While activity values are usually available at the international/national level, this is not the case at regional/local scales, where only emission inventories (PRESSURES) are compiled.

A further key issue for future research is related to activity evolution. On the one side, one would certainly like to improve the estimation of how local economic sectors will develop and adapt in the future, taking into account both internal factors, such as economic downturns, and external ones, as climate changes. This means considering new land use policies (activity location) as part of the IAM problem. On the other side, since a perfect prediction of activity evolution is out of question, new methods to deal with uncertain predictions (ensemble modelling, risk aversion, …) have to be developed and possibly become standard.

4.3.2 Pressures (Emissions)

In the IAM database collected by APPRAISAL, 70 % of the respondents identified emission values as the main weakness of their modelling approach. Quantifying the effectiveness of specific abatement measures within a zone presumes that the emission inventory is disaggregated with sufficient details both spatially and per categories to properly consider the emission abatement measures. This level of detail is unfortunately lacking in most inventories leading to uncertain estimates of the effect of measures. The official national and European (EMEP) emission inventories only contain emission totals for the member state as a whole (or alternatively, gridded data with only SNAP macro-sector detail). Almost all studies focusing on local/urban scales identify, as a major issue, the lack of comprehensive, accurate and up-to-date emission data from bottom-up emission estimation methods. Relevant information on desirable practice for compiling such local emission inventories can be found in the guidelines of the FAIRMODE workgroup on 'Urban emissions and Projections' and the report on 'Integrated Urban Emission Inventories' of the Citeair II INTERREG project (http://www.citeair.eu/).

There is a need for general methodologies for emission inventories that allow:

- Consistent harmonization of bottom-up and top-down emission inventories, to allow "seamless" integration of measures from local to EU level, and vice versa;
- Development of approaches to improve the quality of emission inventories, to 'validate' them and to assess the emission level uncertainty (inverse modelling, source-apportionment methods, new model chains to describe projections, …);
- Adaptation of disaggregation coefficients (in space and time) to regional and local scales, especially for CO, PM and NH_3 emissions.

Additionally, emission projections need to improve data consistency: for instance, the transport sector still lacks data regarding the real vehicle fleet composition (especially the split between different categories of vehicles age and engines type). A finer description for biogenic emissions is also required, better considering data on land use, meteorology and topography (slopes and orientation), according to the species, which can effectively be taken into account, particularly in mountainous and coastal areas.

Emissions factors are another critical point that deserves deeper consideration in particular to define PM components (e.g. BC, metal, UFP, wildfires) and allow to compute the emission of other gaseous pollutants (VOC, SLCP, reactive nitrogen), of HFC emissions from refrigeration and air conditioning equipment's and NO_2 emissions from catalytic converters of cars, as well as those resulting from agricultural fertilizers. Another important improvement would be the splitting of aggregated road traffic emission factors to account for the continuous changes and evolutions of the real vehicle fleets at local, regional and higher levels.

4.3.3 State (Concentration Levels)

Key areas to be addressed by research and innovation in the STATE module refer to both actual measurements and modelling tools.

From the point of view of measurements, we suggest to develop a stronger integration of ground-based and remote-sensing monitoring methods, to assess the "current" AQ situation at a wider scale as well as improve the understanding of the composition of the various PM fractions.

As to models, in order to better assess the AQ state (and the associated health impacts), research should be oriented to better represent AQ at a very detailed scale. This could be done either through the use of Computational Fluid Dynamics (CFD) to explicitly represent local and street levels or by developing sub-grid scale models and parameterization within Chemical Transport Models. Concerning meteorological models, a better use of urban modules in mesoscale models would benefit to regional and more local studies, and help to link models at different scales.

Modelling the urban or local scales requires the inclusion of specific small-scale processes, but also to consider the influence of larger scale effects. This is a challenge that still needs to be worked on because common practices are mainly based on the application of mesoscale models to urban areas without the proper urban parameterizations, and on Gaussian models that are still limited, even with the latest developments.

The use of CFD models to simulate urban areas, forced by a mesoscale model, is a current research area, still with strong limitations because of the high demand of computer time. It is still impossible to simulate a full year period with this modelling approach without several simplifying assumptions. In the future, these limitations could be overcome and the development of the proper link between the mesoscale and the CFD models should therefore be considered as a key research area.

This said, there are still some processes that require a better description within the models. In general, air quality models tend to underestimate peak PM concentrations while exceedances for PM are often considered the most meaningful index in terms of health impact. Further research is required to improve modules for describing windblown dust, resuspension and the formation and fate of secondary organic aerosols. Significant scientific uncertainties also remain regarding the relative contributions of the major components of fine PM, especially organic carbon and metals/dust. In particular, substantial uncertainties in gas-phase and aqueous-phase chemistry mechanisms remain, including key inorganic reactions, aromatic and biogenic reactions and aqueous-phase chemistry. Future research might also include stratospheric chemistry as the spatial domain for air quality models increases when climate applications are considered. The exchange processes with the surface should be further improved considering for example surface bidirectional exchange (ammonia, mercury or polyaromatic hydrocarbons) or the interaction with vegetation, and models have to better couple physics (meteorology) and chemistry processes. This is not only relevant for connecting air quality and climate change modelling, but it is also important when moving to smaller scales (<1 km) where the meteorological models start to resolve turbulent eddies.

Measurements contain valuable information, which can be used as complementary input to modelling results. It is striking that in 40 % of the APPRAISAL reported studies, measurement data were not used at all, not even for model evaluation. This is clearly a point where air quality assessment reports and more specifically air quality plans could be improved. Even if affected by an intrinsic imprecision, monitoring data have the clear advantage that field concentration levels are evaluated with much more accuracy than model results. The main question, which arises in IA applications, is: "how these measurement data can be used most appropriately?" Most of the model results in IA studies are dealing with future projections under certain policy options. By definition, no measurement data are available for this kind of future estimates. A key approach to this problem is to use measurement data in combination with model results at least for the reference case of a recent year. This reference case is most often used as a starting point in the IA exercise. This procedure is referred to as "model calibration" or "data assimilation".

Discussion arises when this combined information has to be used for the simulation of policy scenarios. The use of data assimilation corrections (or calibration factors) as "relevant" information for scenario runs is generally considered appropriate. However, specific and well-defined methodologies to do so are not at hand. One possible approach is to assess the simulated concentration changes of a set of specific policy options in relation to the reference case/year. The resulting concentration changes (so called "deltas") can then be applied on top of the calibrated or data assimilated concentration fields of the reference year (see for example Kiesewetter et al. 2013). However, more research is required to pin down appropriate methodologies to combine reference year measurements with modelling results for future policy scenarios.

Model evaluation is inherent to all these developments and also to common modelling practice. There are already several reported and applied procedures to evaluate models (including model intercomparison exercises), but with different purposes and focusing on particular types of models and/or applications. There is enough information to provide a standardized evaluation protocol organized according to different modelling needs and characteristics. This protocol would be particularly important for stakeholders who need to understand model results in order to decide and implement air quality improvement measures. FAIRMODE activities are addressing this challenge, but a stronger focus on the urban and local scales is needed.

Optimization problems cannot embed full 3D deterministic multi-phase models for describing the nonlinear dynamics linking precursor emissions to air pollutant concentrations because of their computational requirements. IAMs therefore rely on simplified relationships for describing the links between emissions and air quality, which are called "source/receptor (S/R) relationships" (or "surrogate models"). These types of models can be both linear and nonlinear, and examples can be found in literature for both types of approaches. Future research will need to extend surrogate model approaches to properly describe the most important processes in terms of chemistry, meteorology at the appropriate scale accounting for potential non-linearity. Moreover, it will need to focus on proper "Design of Experiments" methods (that is to say, the way in which CTM simulations should be planned, for identification of the surrogate models). On the one hand they need to maximize the information used to identify S/R relationships and, on the other hand, to limit the number of CTM simulations required to derive these relationships.

Finally, integrated assessment long-term studies should take into account both air quality and climate change issues. In this framework, it is important to develop the use of future meteorological simulations for running AQ models. A challenge is the development in IAM of online chemical transport models, which allow the study of feedback interactions between meteorological/chemical processes within the atmosphere, and thus take into account AQ/climate change connections.

4.3.4 Impact (Human Health)

Traditionally, modelling tools have addressed air quality assessment issues including dispersion and chemistry but rarely have considered also exposure or health indicators. However, Health Impact Assessment (HIA) should be part of integrated assessment, as it usually involves a combination of procedures, methods and tools by which an air quality policy can be judged in terms of societal impact. Quantification of health effects in HIA (Pope and Dockery 2006) is particularly important, as knowing the size of an effect helps decision makers to distinguish between the details and the main issues that need to be addressed and facilitates decision making by clarifying the trade-offs that may be entailed. Secondly, adding up all positive and negative health effects using appropriate modelling methods allows for the use of economic instruments such as cost-effectiveness analysis, which further aids decision-making.

Exposure-response functions (which quantify the change in population health due to a given exposure) are identified as the main sources of uncertainty in an integrated assessment (Tainio 2009), but it is also important to further explore the "complete individual exposure to air pollution" pathways. "Complete" here means indoor as well as outdoor air pollution over a 24 h/24 h period; "Individual" means monitoring air quality at the person level, possibly using portable and easy-to-wear monitors. These two factors, together with a dynamic view of exposition variations, will result in a more comprehensive view on individual exposure. If this could be combined with human biomonitoring, i.e. measuring the concentration of a certain pollutant or one of its by-products in the human body, it would enrich our current knowledge regarding the impact of air pollution on human health. This would clearly necessitate the consideration of dynamic maps of population and pollution (i.e. considering the hourly population living/working habits depending on age, gender, activity… and modelling air quality maps with the same level of detail).

Most plans and projects are focused on long-term exposure that has much greater public health impact. Not all acute effects are included in long-term impacts and therefore short-term impact on morbidity and mortality might be underestimated. Mortality and morbidity factors of long-term NO_2 and O_3 exposure should also be investigated, as well as NO_2 exposure effects in particularly polluted environments (i.e. busy roads).

Overall, the most critical element in respect of HIA is the lack of general methods to deal with the multi-pollutant case. In all urban areas, in fact, citizens are exposed to a cocktail of different pollutants, the combined effect of which is largely unknown.

4.3.5 Responses (Methodologies to Design Measures)

The RESPONSES module includes methodologies that can be formalized and implemented to design AQ plans. This is related on one side to the type of decisions that can be assumed at local level and how they can be integrated into other policy domains (decision variables), on the other side to the methodologies to select such decisions (decision problem). It is clear that the two aspects are strictly interrelated, and, for instance, the definition of the decision variables can affect the formalization of the decision problem.

As to the first aspect, the inclusion of socio-economic aspects in the decision problem formulation (e.g. the public acceptance of different measures) and the land planning aspect should be considered in AQ plans. Such plans should also be tightly connected with other policy areas (e.g. energy, transport, etc.) and related plans.

Possibly, the main challenge in this field is the inclusion of "non-technical/efficiency measures" within the planning options. The use of these measures is now limited to scenario analysis, because it is very difficult to estimate removal efficiencies and costs of such measures, particularly, because they impact many other sectors beside air quality. For instance, car sharing has the potential to reduce not only exhaust emissions, but also accidents and noise. How can the overall cost be associated to the benefits in such diverse sectors? An additional complexity is related to the use of these measures in an optimization framework; from this point of view, new formal approaches need to be devised.

As to the problem formulation, one major area of investigation for the future is the consideration of dynamic evolution of the physical, economic, and social environment. All current approaches are static, in the sense that they devise a solution to be reached within a given time horizon (say, for instance, in 2020). However, the system we want to control is non-stationary (e.g. the effect of the current economic crisis) and it may therefore be more supportive for decision makers to know where and when to currently invest with the highest priority in order to follow a certain path to the target with the ability to adapt decisions with time, in case the system evolution differs from the projected one. This involves the necessity of flexibly adding into the plans the advent of new technologies and the ability to determine the cost of scrapping old plants to substitute them with newer ones. This essentially means designing a new generation of Decision Support Systems to be intended more as control dashboards, than planning tools. Related to the dynamic problem is also the issue of how to evaluate future benefits of air quality investments. If economy has defined since long how to account for investment costs lasting for a given period in the future, this is more difficult for benefits that are not monetizable or last in the future for an unknown period. How can we account for a 20 % improvement of an air quality index ten years from now? What is the benefit from a reduction of PM10 today that will decrease cardiovascular problems in a population sometime in the future?

A more synergic use of Source Apportionment and Optimization approaches should also be fostered. SA could limit the degrees of freedom of cost-effectiveness analysis, constraining the optimal solution to consider only a subset of the possible measures previously identified applying SA. On the other hand, the optimization approaches can automatically perform source apportionment establishing the most cost-effective emission reductions and identifying the sources categories associated to these reductions, without the need to monitor and chemically characterize air pollutants.

4.4 Areas for Future Research of IAM Systems

A number of directions for future research have been identified by considering the IAM as a whole, in particular related to the integration of IAM scales and the uncertainty assessment.

One point is certainly the development of methodologies integrating widely used source-apportionment and modelling approaches to quantify the effective potential of regional-local policies and of European/national ones in a specific domain.

Different models are designed and implemented to approach different spatial scales (from regional, to local, to street level). Future research should study how to link these different scales and to build an IAM system able to connect different "scale-dependent" approaches consistently, to model policy options from regional, to local, to street scale.

As it is already done with CTMs, a research direction could be devoted to developing IAMs nesting capabilities (both one-way and two-way nesting) to easily manage EU/national constraints at regional level, and at the same time to provide feedbacks from the regional to the EU/national scale.

At the moment, national climate change policies simply dictate some constraints to local air quality plans, but it is well known that also local air quality policies (e.g. the reduction of aerosols) can have consequences in terms of climate change. In a "resource limited" world, the aspect of maximizing the efficiency of the actions (to get win-win solutions for AQ and CC) will become of extreme importance and will require guidelines to integrate climate change policies (normally established at national or even international levels) with air quality plans developed at regional/local level.

Uncertainty estimates are an essential element of integrated assessment, as a whole. Uncertainty information is not intended to directly dispute the validity of the assessment estimates, but to help prioritize efforts to improve the accuracy of those assessments in the future, guiding decisions on methodological choices with respect to the tools that are being used.

In order to assess the total uncertainty and evaluate the performance of an IAM system, the uncertainty related to the different modelling components of the system (meteorological modelling, air quality modelling, exposure modelling, cost-benefit modelling) has to be quantified separately. In literature, there are very few works concerning the application of uncertainty/sensitivity analysis in the IAM considered as a whole system. The most complete works in this frame are due to Uusitalo et al. (2015), who present a quite complete methodological review concerning possible application of uncertainty and sensitivity analysis in IAM, and to Oxley and ApSimon (2007), who reviewed the issues related to uncertainty in IAM, particularly focusing on space and time resolution and on the problem of uncertainty propagation in integrated system. More in general, all works, possibly with the only exception of Freeman et al. (1986), use a numerical approach based on Monte Carlo simulation at different levels of complexity. This is probably due to the increasing computational capacity and to the relatively newness of the problem treatment in the context, causing scientist to directly start the study from the numerical approaches both for uncertainty and sensitivity analysis.

As the chemical and physical processes involved are not linear, and some uncertainties may compensate each other (Carnevale et al. 2016), the interconnection of all IAMs individual uncertainties remains a scientific challenge. Combining all uncertainties to calculate a total uncertainty would require a great number of simulations, accounting for all possible combinations. This complexity does not allow for setting straightforward quality criteria in terms of IAMs, even though IAM is considered an important policy tool.

In more detail, some of the issues still to be investigated on IAM concern:

- The optimization algorithms. The decision problem is solved by means of optimization algorithms. How does the optimization algorithm bias the determination of effective policies?

- The planning indicators for human, ecosystems and materials exposure. The decision problem determines the abatement measures or other actions that optimize the objectives, and that have to comply with the physical, economic and environmental constraints. Objectives and environmental constraints are typically indicators of human, ecosystems and material exposure. How do different sets of indicators impact on policies design?
- The source/receptor relationships. What is the uncertainty of source/receptor relationships? Which is the sensitivity of the decision problem solutions to different source/receptor relationships?
- The emission and climatic conditions. Such source/receptor relationships are identified processing CTM simulations for different reference years, meaning for some specific emission and meteorological scenarios. The overall results of IAM application are indeed variations with respect to these conditions that probably will not be exactly replicated in the future, when decision will be implemented. How do the assumption about these reference years impact the design of policies?

In general, all these points highlight the need of defining a set of indexes and a methodology to measure the sensitivity of the decision problem solutions. It is in fact worth underlining that, while for air quality models the sensitivity can be measured by referring in one way or the other to field data, for IAMs this is not possible, since an absolute "optimal" policy is not known and most of the times it does not even exist. The traditional concept of model accuracy must thus be replaced by notions such as risk of a certain decision or regret of choosing one policy instead of another. Indeed, since long ago, the "UNECE workshop on uncertainty treatment in integrated assessment modelling" (UNECE 2002), concluded that policy makers are mainly interested in robust strategies. Robustness implies that optimal policies do not significantly change due to changes in the uncertain model elements. Robust strategies should avoid regret investments (no-regret approach) and/or the risk of serious damage (precautionary approach) (Amann et al. 2011).

Acknowledgments This chapter is partly taken from APPRAISAL Deliverable D2.7 (downloadable from the project website http://www.appraisal-fp7.eu/site/documentation/deliverables.html).

References

Amann M, Bertok I, Borken-Kleefeld J, Cofala J, Heyes C, Höglund-Isaksson L, Klimont Z, Nguyen B, Posch M, Rafaj P, Sandler R, Schöpp W, Wagner F, Winiwarter W (2011) Cost-effective control of air quality and greenhouse gases in Europe: modelling and policy applications. Environ Model Softw 26:1489–1501

Borrego C, Sá E, Carvalho A, Sousa J, Miranda AI (2012) Plans and Programmes to improve air quality over Portugal: a numerical modelling approach. Int J Environ Pollut 48(1/2/3/4): 60–68

Builtjes P, Jörß W, Stern R, Theloke J (2010) Particle Reduction Strategies PAREST (Final Report). FKZ 206 43 200/01. Contracting authority: German Federal Environment Agency (UBA). Umweltbundesamt, Dessau-Roßlau (in German)

Carnevale C, Finzi G, Pisoni E, Volta M (2011) Minimizing external indirect health costs due to aerosol population exposure: a case study from Northern Italy. J Environ Manage 92:3136–3142

Carnevale C, Finzi G, Pisoni E, Volta M, Guariso G, Gianfreda R, Maffeis G, Thunis P, White L, Triacchini G (2012) An integrated assessment tool to define effective air quality policies at regional scale. Environ Model Softw 38:306–315

Carnevale C, Douros J, Finzi G, Graff A, Guariso G, Nahorski Z, Pisoni E, Ponche J-L, Real E, Turrini E, Vlachokostas Ch (2016) Uncertainty evaluation in air quality planning decisions: a case study for Northern Italy. Environ Sci Policy, in press

Coll I, Lasry F, Fayet S, Armengaud A, Vautard R (2009) Simulation and evaluation of 2010 emission control scenarios in a Mediterranean area. Atmos Environ 43:4194–4204

Cuvelier C, Thunis P, Vautard R, Amann M, Bessagnet B, Bedogni M, Berkowicz R, Brandt J, Brocheton F, Builtjes P, Carnavale C, Copalle A, Denby B, Douros J, Graf A, Hellmuth O, Honoré C, Hodzic A, Jonson J, Kerschbaumer A, de Leeuw F, Minguzzi E, Moussiopoulos N, Pertot C, Peuch VH, Pirovano G, Rouil L, Sauter F, Schaap M, Stern R, Tarrason L, Vignati E, Volta M, White L, Wind P, Zuber A (2007) CityDelta: a model intercomparison study to explore the impact of emission reductions in European cities in 2010. Atmos Environ 41:189–207

D'Elia I, Bencardino M, Ciancarella L, Contaldi M, Vialetto G (2009) Technical and Non-Technical Measures for air pollution emission reduction: the integrated assessment of the regional Air Quality Management Plans through the Italian national model. Atmos Environ 43:6182–6189

De Ridder K, Lefebre F, Adriaensen S, Arnold U, Beckroege W, Bronner C, Damsgaard O, Dostal I, Dufek J, Hirsch J, IntPanis L, Kotek Z, Ramadier T, Thierry A, Vermoote S, Wania A, Weber C (2008) Simulating the impact of urban sprawl on air quality and population exposure in the German Ruhr area. Part I: reproducing the base state. Atmos Environ 42:7059–7069

Freeman DL, Egami RT, Robinson NF, Watson JG (1986) A method for propagating measurement uncertainties through dispersion models. J Air Pollut Control Assoc 36:246–253

Juda-Rezler K (2004) Risk assessment of airborne sulphur species in Poland. In: Borrego C, Incecik S (eds) Air pollution modelling and its application XVI. Kluwer Academic/Plenum Publishers, Hardbound, pp 19–27

Karvosenoja N, Klimont Z, Tohka A, Johansson M (2007) Cost-effective reduction of fine particulate matter emissions in Finland. Environ Res Lett 2:044002

Kiesewetter G, Borken-Kleefeld J, Schöpp W, Heyes C, Bertok I, Thunis P, Bessagnet B, Terrenoire E, Amann M (2013) TSAP Report #9: Modelling compliance with NO2 and PM10 air quality limit values in the GAINS model. Service Contract on Monitoring and Assessment of Sectorial Implementation Actions (ENV.C.3/SER/201 1/0009)

Lefebvre W, Fierens F, Trimpeneers E, Janssen S, Van de Vel K, Deutsch F, Viaene P, Vankerkom J, Dumont G, Vanpoucke C, Mensink C, Peelaerts W, Vliegen J (2011) Modeling the effects of a speed limit reduction on traffic-related elemental carbon (EC) concentrations and population exposure to EC. Atmos Environ 45:197–207

Mediavilla-Sahagun A, ApSimon H (2013) Urban scale integrated assessment of options to reduce PM10 in London towards attainment of air quality objectives. Atmos Environ 37:4651–4665

Oxley T, ApSimon H (2007) Space, time and nesting integrated assessment models. Environ Model Softw 22:1732–1749

Oxley T, ApSimon H, Dore A, Sutton M, Hall J, Heywood E, Gonzales del Campo T, Warren R (2003) The UK Integrated Assessment Model, UKIAM: a national scale approach to the analysis of strategies for abatement of atmospheric pollutants under the convention on long-range transboundary air pollution. Integr Assess 4:236–249

Pope C, Dockery D (2006) Health effects of fine particulate air pollution: lines that connect. J Air Waste Manag Assoc 56(6):709–42

Syri S, Karvosenoja N, Lehtilä A, Laurila T, Lindfors V, Tuovinen J-P (2002) Modeling the impacts of the Finnish Climate Strategy on air pollution. Atmos Environ 36:3059–3069

Tainio M (2009) Methods and uncertainties in the assessment of the health effects of fine particulate matter (PM2.5) air pollution. National Institute for Health and Welfare (THL) and Faculty of Natural and Environmental Science at Kuopio University

Thunis P, Rouil L, Cuvelier C, Stern R, Kerschbaumer A, Bessagnet B, Schaap M, Builtjes P, Tarrason L, Douros J, Moussiopoulos N, Pirovano G, Bedogni M (2007) Analysis of model responses to emission-reduction scenarios within the CityDelta project. Atmos Environ 41:208–220

UNECE (2002) Progress report prepared by the Chairman of the task force on integrated assessment modelling, United Nations Economic Commission for Europe, Geneva, Switzerland

Uusitalo L, Lehikoinen A, Helle I, Myrberg K (2015) An overview of methods to evaluate uncertainty of deterministic model in decision support. Environ Model Softw 63:24–31

Van Jaarsveld JA (2004) The operational priority substances model. Report nr 500045001/2004, RIVM, Bilthoven, The Netherlands

Vlachokostas Ch, Achillas Ch, Moussiopoulos N, Hourdakis E, Tsilingiridis G, Ntziachristos L, Banias G, Stavrakakis N, Sidiropoulos C (2009) Decision support system for the evaluation of urban air pollution control options: application for particulate pollution in Thessaloniki, Greece. Sci Total Environ 407:5937–5938

Zachary DS, Drouet L, Leopold U, Aleluia Reis L (2011) Trade-offs between energy cost and health impact in a regional coupled energy–air quality model: the LEAQ model. Environ Res Lett 6:1–9

Chapter 5
Two Illustrative Examples: Brussels and Porto

C. Carnevale, F. Ferrari, R. Gianfreda, G. Guariso, S. Janssen, G. Maffeis, A.I. Miranda, A. Pederzoli, H. Relvas, P. Thunis, E. Turrini, P. Viaene, P. Valkering and M. Volta

5.1 Introduction

To evaluate in practice how IAM can be used to formulate and improve current air quality plans, this chapter reports on the application of one of the existing IAM tools, to two test cases: one is the Brussels Capital Region in Belgium and the other the region of Porto in the North of Portugal. The two cases are representative for the two options that are available for the decision pathway in the IAM framework as presented in Chap. 2: the scenario evaluation and the optimisation. Before presenting the peculiarities and the results obtained for the two test cases, this chapter briefly describes the specific features of the IAM tool used, namely RIAT+.

C. Carnevale · A. Pederzoli · E. Turrini · M. Volta (✉)
Università degli Studi di Brescia, Brescia, Italy
e-mail: marialuisa.volta@unibs.it

F. Ferrari · R. Gianfreda · G. Maffeis
Terraria srl, Milan, Italy

G. Guariso
Politecnico di Milano, Milan, Italy

S. Janssen · P. Viaene · P. Valkering
Vlaamse Instelling voor Technologisch Onderzoek n.v. (VITO), Mol, Belgium

A.I. Miranda · H. Relvas
Universidade de Aveiro, Aveiro, Portugal

P. Thunis
European Commission, Joint Research Centre (JRC), Directorate for Energy, Transport and Climate, Air and Climate Unit, Ispra, Italy

G. Guariso and M. Volta (eds.), *Air Quality Integrated Assessment*,
PoliMI SpringerBriefs, DOI 10.1007/978-3-319-33349-6_5

5.2 The RIAT+ System

The RIAT+ system was developed during the EU OPERA project (www.operatool.eu) and it is intended to help regional decision makers select optimal air pollution reduction policies that will improve the air quality at minimal (industrial and/or external) costs. To achieve this, the system incorporates explicitly the specific features of the area of interest with regional input datasets for the:

- Precursor emissions of local and surrounding sources;
- Abatement measures (technical and non-technical) described per activity sector and technology with information on application rates, emission removal efficiency and cost;
- The effect of meteorology and prevailing chemical regimes through the use of site-specific source/receptor functions.

The system runs as a stand-alone desktop application and can be downloaded from the OPERA project website (http://www.operatool.eu/download/). The package is distributed with a personal, non-exclusive and royalty-free license and has been applied in various regions, such as Emilia-Romagna (Carnevale et al. 2012) and in Alsace (Carnevale et al. 2014).

The RIAT+ software implements both the possible decision pathways introduced in Chap. 2 in the light of the classical DPSIR (Drivers-Pressures-State-Impact-Responses) scheme, adopted by the EU:

- **Scenario analysis**, where emission reduction measures are selected on the basis of expert judgment or Source Apportionment and then tested through simulations of an air pollution model.
- **Optimisation**, where the set of cost effective measures for air quality improvement are automatically selected by solving a multi-objective optimization problem.

To allow both approaches to be implemented in a fast and handy way (i.e. to be able to support a real-world discussion about possible options) a key feature of RIAT+ is a S/R model used to relate emissions (pressures) to a suitable air quality indicator, AQI (state). In principle, such a S/R model should be a full complex chemical transport model, but in practice this would be impossible for the computational burden that such models imply. So, within RIAT+, the relations between emissions and air quality indicators are expressed by means of Artificial Neural Networks (ANNs), that, in turn, are tuned to replicate the results of deterministic air quality model. ANNs are often referred to in this context as "surrogate models". The reason for this choice is that neural networks are known to be suitable to describe a nonlinear relationship between data, such as those theoretically involved in the formation of air pollution. Their identification procedure requires two steps: (1) the definition of the specific structure, and (2) the calibration of the parameters to the specific application. The selected structure of the ANNs must be able to retain what are considered to be the essential features of the original model. As the value

Fig. 5.1 *Quadrant shape* input configuration

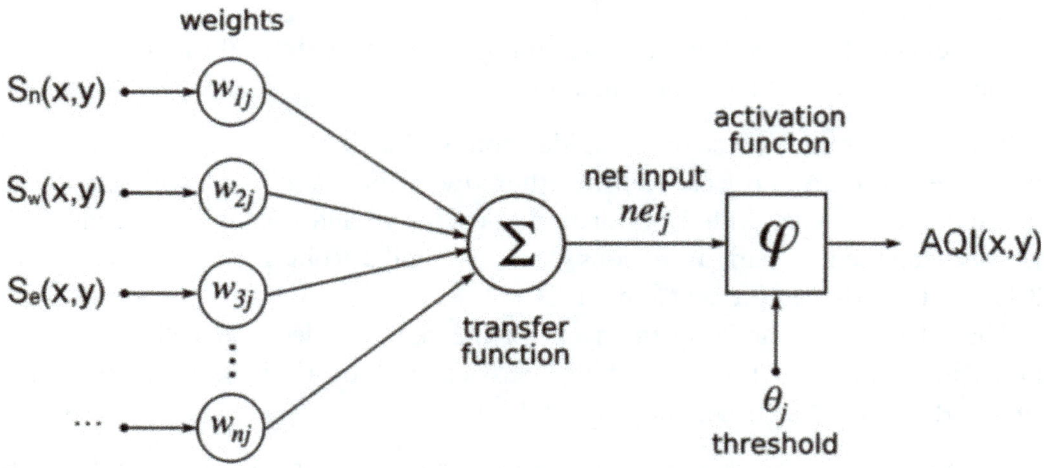

Fig. 5.2 A sketch of an elementary neuron

of an air quality index is not only dependent on the local precursor emissions but also on surrounding emissions, the surrogate models must consider the influence of these surrounding emissions and the prevalent wind direction. This is achieved by considering a quadrant shape input configuration as shown in Fig. 5.1 where the emissions $S_j(x, y)$ are summed according to these quadrants, the dimension of which depends on the specific area and pollutant under study, and then used to compute the AQI value in position (x, y). Such a calculation is performed using a network of connected elements (neurons), the structure of which is sketched in Fig. 5.2.

The development of the surrogate models thus means: first, the definition of the input variables and of the form of the so-called "activation function" φ, generally a strongly nonlinear function of its argument, which is in turn a weighted sum of the input values; second, the determination of all the model parameters (namely, the weights w_{ij} and the threshold θ_j).

This second step (training) is performed by imposing that the surrogate model represent, as much as possible, a set of CTM calculations that are representative of the range of emissions/AQI that may be entailed by the plan to be developed. The process of selecting such configurations to be simulated by the CTM is usually

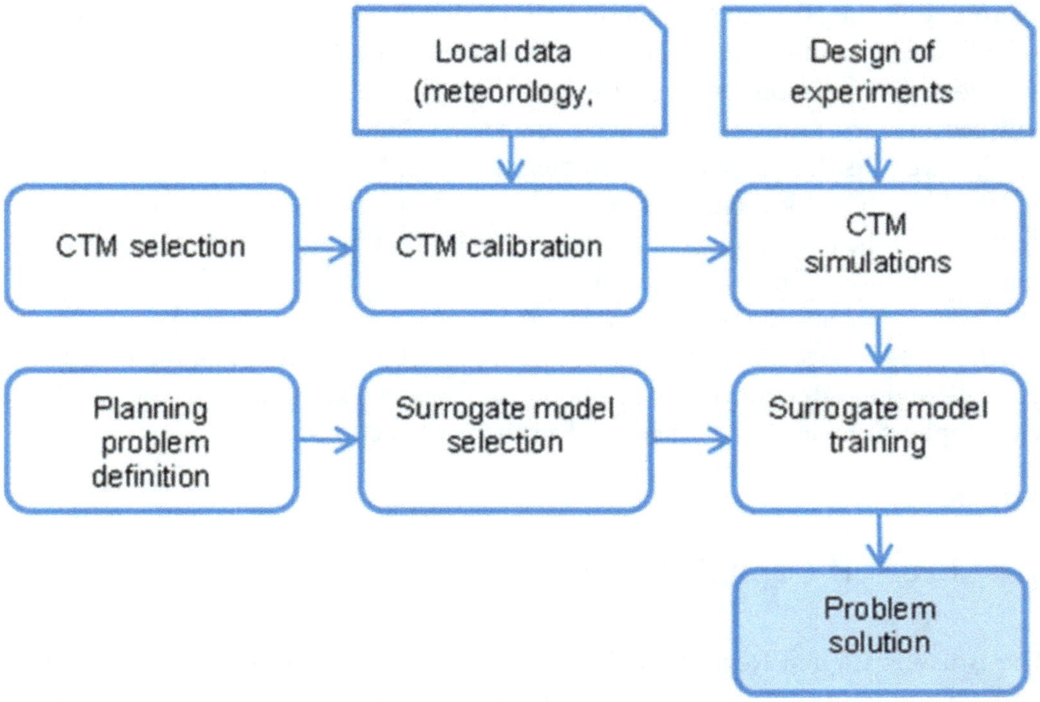

Fig. 5.3 RIAT+ solution procedure

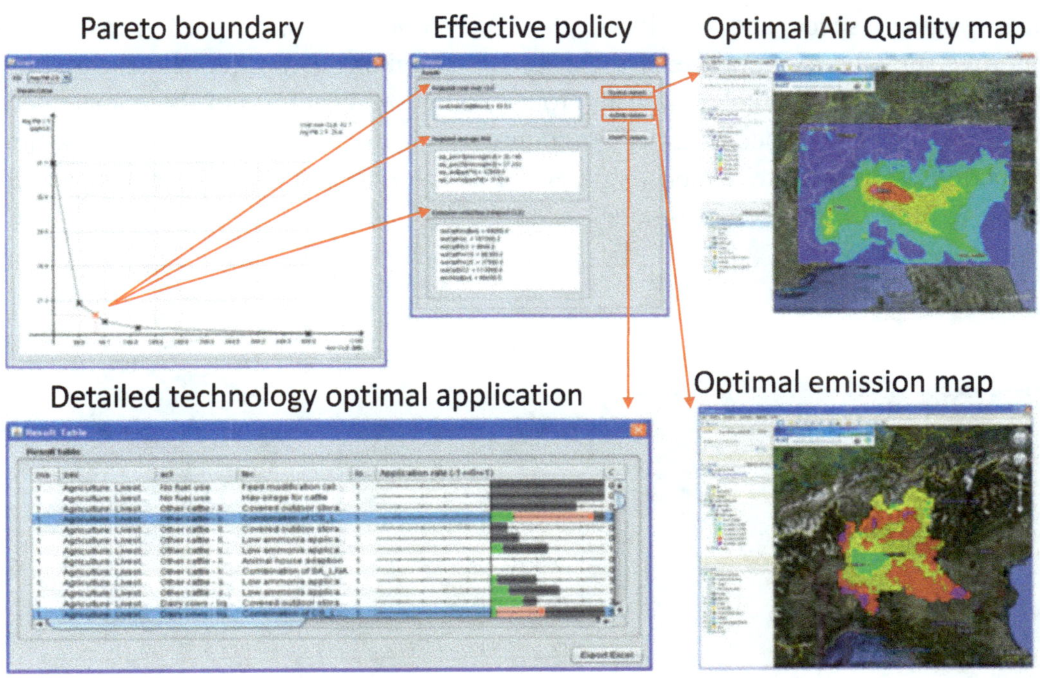

Fig. 5.4 RIAT+ output

referred to as the 'Design of Experiment'. On the one hand, these simulations have to be limited in number due to their computational time, but, on the other hand, they must be able to represent as closely as possible the cause-effect relationship between precursor emissions and the considered AQIs.

The overall solution procedure implemented in RIAT+ is presented in Fig. 5.3, which shows how local data, CTM simulation and problem statement are combined to determine the overall results. These can in turn be analysed under different views: values and costs of measures in different sectors, spatial distribution of emissions and AQIs, efficient trade-offs between costs and AQIs (see Fig. 5.4).

RIAT+ IAM system has been used in support of air quality planning in Brussels Capital Region (Belgium) and in the Great Porto Area (Portugal). The results of such applications are briefly sketched in the next sections.

5.3 Brussels Capital Region

The Brussels Capital Region (BCR) has an area of 161 km^2 and is home to more than 1.1 million people. The region consists of 19 municipalities, one of which is the Brussels Municipality, the capital of Belgium. The location of the BCR in Belgium is shown in Fig. 5.5.

For the BCR, Brussels Environment, BIM (http://www.ibgebim.be) is responsible for the study, monitoring and management of air, water, soil, waste, noise and nature (green space and biodiversity). BIM proposed a list of 13 measures to improve air quality, approved by the Brussels authorities and consisting of nine measures related to vehicle traffic and four to domestic heating. For these abatement measures, BIM provided order-of-magnitude estimations of the costs and emission reductions. These were screened to determine the effect of the different measures using RIAT+ in the scenario mode.

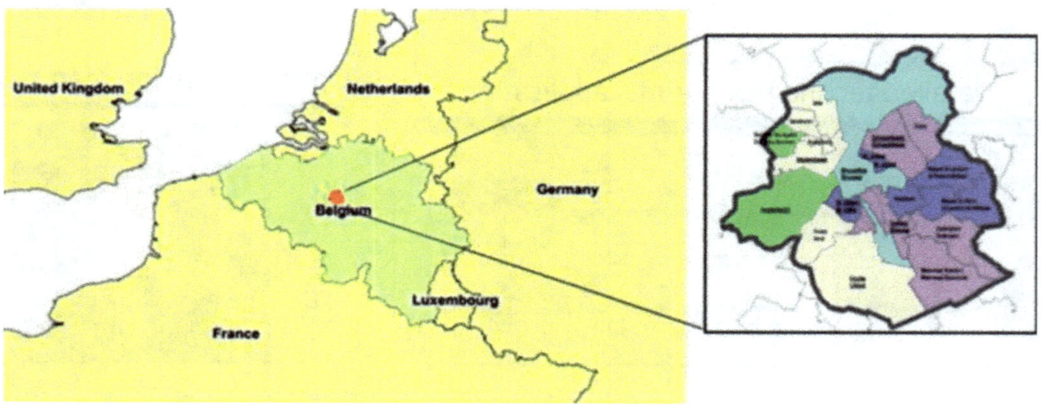

Fig. 5.5 Location of the BCR (*dark area*) in Belgium

5.3.1 Proposed Abatement Measures

Traffic

Reduction measures related to traffic are mainly non-technical, meaning that they involve a change of the traffic flows more than a change in the emission standards. The main actions that were analysed are:

- The introduction of a low emission zone (LEZ) extended to the entire capital region or only the inner part of Brussels municipality. Two possibilities were tested: a restriction only for Heavy Duty Vehicles (HDV) with emission standard prior to Euro 5; or a restriction also for passenger cars with diesel engine before Euro 5 and gasoline before Euro 2. Emission and cost data related to these cases were derived from a TM-Leuven (2011) study. Reduction entailed by these plans ranged, for instance for PM2.5, between 10 and 40 % with respect to the CLE scenario.
- Reduction of the car parking lots available in Brussels by 25,000. This measure is assumed to reduce the number of commuters entering Brussels every day and discourage the local inhabitants from using cars to drive to work. The estimate of BIM/IBGE (2012) is that 140,000 commuters enter Brussels in every weekday and 225,000 residents use their vehicle to get to work. Given that the estimated distance travelled is about 9 km for residents and 13 km for the others, and assuming that the reduction of parking places entails a corresponding reduction of trips, this measure would mean a reduction of 129 M km a year with a 1.5 % reduction of the traffic emissions.
- The implementation of mobility plans encouraging public transport for all the sites hosting more than 1000 people and all events involving more than 1000 participants. This is assumed to equal a 3.7 % reduction of the traffic sector, with a correspondent reduction, for instance, of 2.6 % of PM10 emissions.
- A modal shift from car to bike for commuting. This follows existent plans to move from the current 1.9 % of commuting trips by bike to 20 % by 2018. An English study indeed showed that each new cyclist corresponds to a 500 € gain per year for society, mainly through the reduction of costs in health care (Cycling England 2007). This would correspond to a further reduction of commuting private traffic by 4.8 %.
- The introduction of a urban toll. This can be implemented according to different schemes: a toll of 12 €/day within BCR; one of 3 €/day in the larger Brussels Regional Express Network (RER); a price of 7 c€/km in the RER zone. The first scheme is estimated to reduce NOx emission by 18 % with respect to CLE 2018, the second by 11 %, the third by 9 %. According to a STRATEC (2014) study, the net present value of the costs for implementing these scheme ranges from 250 M€ for the first two, to about 2500 M€ for the third.
- Eco-driving. To make eco-driving the standard on roads, it should be first taught during the various formations of the road users (driving license, taxi driver permit, training of bus and truck drivers, etc.); but we must also regularly sensitize drivers by information and awareness tools, particularly within

enterprise transport plans. Following AIRPARIF (2012), it is assumed that about 25 % of all drivers are susceptible to a more eco-driving style, implying 7 % less fuel use, and hence a 1.7 % reduction of emissions. Assuming a full scale eco-driving campaign similar to that implemented in the Netherlands (ECODRIVEN 2008) will result in a rough estimate of 180 k€ annually. This implies a net present value, discounted over a time period of 6 years (e.g. 2014–2020) on the basis of a 5.7 % interest rate, of about 1 M€.

- Stimulating the use of Compressed Natural Gas (CNG) as car fuel. While this is a more technical measure, it seems that in Belgium is more a psychological problem than a lack of infrastructure. It is necessary to implement incentives and information campaigns and to increase the number of points of sale sufficiently to make CNG a viable alternative as in many other countries. With respect to the 2010 situation (FEBIAC 2013), it was assumed that 540,000 (10 % of the fleet) could run on CNG in 2020, substituting 6.3 % of current diesel cars and 3.7 % of gasoline cars. Since the average mileage is 15,500 km/year, this would mean a decrease of 7.6 % of NOx emitted by cars.

Domestic heating

The reduction of emissions in the thermal energy sector was planned along both technical and non-technical actions.

- Maintenance of residential heating appliances. This measure consists of a periodic inspection of boilers, according to the requirements listed in the PEB (Performance Energétique des Bâtiments) guidelines. In particular, in this study, the measure was only applied to residential boilers, with a power in excess of 20 kW, which corresponds, to 95 % of all boilers in the residential sector. Specifically, the periodic inspection of boilers consists of cleaning all components of the boiler and flue system, the burner setting and compliance verification requirements. Oil-fired boilers should be checked annually while natural gas boilers should be checked every three years. The adoption of this measure is assumed to reduce NOx emissions by 72 ton, SOx by 33 ton, VOC by 9.3 ton, and PM2.5 by 3 ton in 2020. According to the estimate of VITO (2011) the adoption of this action may cost around 18 M€/year.
- Improving building isolation. This measure aims to stimulate the construction and building renovation programs by demonstrating that it is possible to achieve excellent energy and environmental performance while opting for economically justifiable solutions and promoting high architectural quality. It provides building owners the opportunity to be ambitious, and allows to generate a number of exemplary buildings that have a lasting effect on the Brussels construction market through the experience obtained. Between 2007 and 2013, 520,000 m^2 of buildings were renovated with an improved isolation in the Brussels area.
- Local plan for energy management and energy audits. This measure is mainly constituted by an analysis of existing or renovated building owned by large real estate companies to define where energy-saving maintenance is needed, and is

mandatory under city government regulations. It is estimated to reduce NOx emission by about 24 ton in 2020, VOC emissions by slightly more than 3 tons and only by 0.6 ton PM2.5. The cost has not being evaluated since they are part of CLE 2020.

5.3.2 The 2010 Scenario

The starting situation for the simulations was a reconstruction of 2010 situation based on the emission inventory of the previous year. For the two sectors involved, namely domestic heating (SNAP code 2) and traffic (SNAP 7) the emissions are listed in Table 5.1.

Table 5.2 reports all the reductions per each pollutant and each measure that can be obtained by their full adoption.

The air quality modelling system AURORA (Mensink et al. 2001; Lauwaet et al. 2013) was used in Brussels capital region to simulate the transport, chemical transformations and deposition of atmospheric constituents at the urban to regional scale. It consists of several modules. The emission generator calculates hourly pollutant emissions at the desired resolution, based on available emission data and proxy data to allow for proper downscaling of coarse data. The actual CTM then uses hourly meteorological input data and emission data to predict the dynamic behaviour of air pollutants over the study area. This results in hourly three-dimensional concentration and two-dimensional deposition fields for all species of interest. For the BCR, AURORA was set up for a domain of 49 $\times$ 49 grid cells at 1 km resolution. For the vertical discretisation, 20 layers were used for a domain extending up to 5 km. The layer thickness increases from 27 m for the bottom layer to 743 m for the top layer. For the boundary conditions, the results of an AURORA run were used for a domain covering Belgium at a resolution of 4 km. The same boundary conditions were used in all runs. For the meteorological inputs, the ECMWF ERA INTERIM data with a resolution of 0.25° were used and interpolated to the model grid. The emissions are based on the CORINAIR emission inventory, which were spatially disaggregated using the Emission MAPping tool (E-MAP) developed by VITO (Maes et al. 2009). This tool downscales national emission inventories using a set of proxy data, such as land use information or the road network. The carbon bond CB05 chemical mechanism (Yarwood et al. 2005) was used.

Table 5.1 Emissions (ton/year) in the BCR for the base scenario

Sector	NOx	CO	SOx	VOC	PM2.5
Domestic heating	2266	3899	586	299	71
Traffic	2026	1581	5	5	130

Table 5.2 Emissions reduction (%) wrt the base scenario

Measure description	Emission reduction (%)				
	NOx	SOx	VOC	PM2.5	PM10
Eco driving	1.7	1.7	1.7	1.7	1.7
Modal shift	2.4	0.0	2.4	3.4	3.4
Transport plans	1.9	0.0	1.9	2.6	2.6
Urban toll	18	19	23	15	15
Parking places	0.8	0.0	0.8	1.1	1.1
Boiler maintenance	3.2	5.6	3.1	4.2	4.2
Exemplary buildings	0.2	0.0	0.2	0.1	0.1
Energy efficiency large bldgs	0.3	0.3	0.3	0.3	0.3
Energy audits	1.4	1.1	1.5	1.0	1.0

The model results were validated by comparison to the observed values at the measurement stations inside the domain. The methodology proposed by FAIRMODE (see: Thunis et al. 2012, 2013; Pernigotti et al. 2013) was adopted. Briefly put, the methodology accounts for observation uncertainty in the evaluation of model results and proposes a method to decide whether model results are acceptable. In Fig. 5.6 the target diagram for the NO_2 results is shown. For the results to be acceptable the method requires that they lie within the unit circle. In this sense, evaluating a target diagram is much the same as looking at a darts board, the aim being to have all points as close to the centre as possible. In the present case, all stations are well within the unit circle, except for the single point corresponding to a suburban traffic station where the underestimation of observed values (bias) is bigger although still within the acceptable range. The correspondent diagram for PM10 shows a slightly worse model performance.

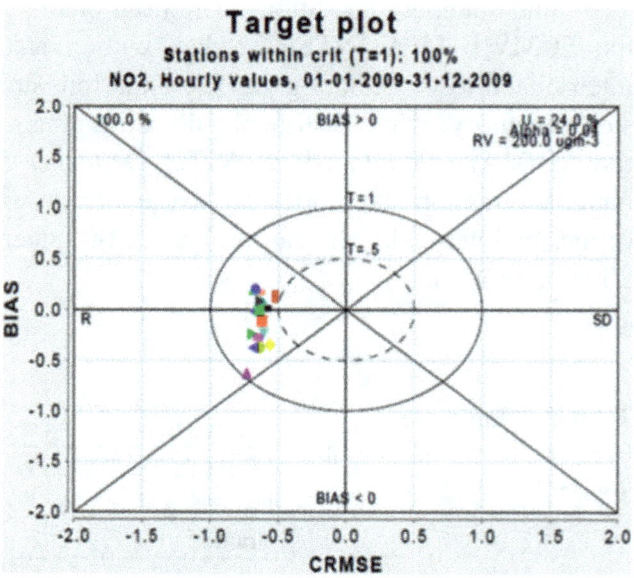

Fig. 5.6 Validation analysis of AURORA model results for NO_2. The markers correspond to the situation of 18 measurement stations (CMRSE is centered root-mean-square error)

5.3.3 Design of Experiments and Source/Receptor Models

For the Design of Experiment phase, three levels of emission within BCR were distinguished: base case (B), high emission reductions (H) and low emission reductions (L). They correspond to the following assumptions:

- The B emission level corresponds to the CLE2020 emissions, increased by 20 %.
- The H level is obtained by projecting the 2009 regional emission inventory to 2020 with the projected application rates of all technologies (as predicted by the GAINS inventory).
- The L level (low emission reductions) is obtained as the average of B and H.

The emission levels for the model grid cells outside the BCR were also changed according to the changes inside the BCR, while for the boundary conditions of the 49 × 49 km domain, the average emission variation from the 2009 inventory projected to CLE2020 of the BCR domain cells was applied to the emission inventory covering all Belgium.

To determine the emission reduction scenarios for the ANN training, the three levels B, H, L were combined to produce 14 different emission sets.

The selected Air Quality Indexes (AQIs) were: yearly average of PM10 concentrations and yearly average of NO_2 concentrations. The emissions surrounding individual model grid cells were aggregated according to the quadrants in Fig. 5.1 with a dimension of 14 cells for PM10 and of 20 cells for NO_2.

The ANNs ability to reproduce CTM results was checked in different ways. For instance, Fig. 5.7 shows the results of the differences between other scenarios and the low reduction case.

The ANN is well capable of reproducing the CTM behaviour for NO_2 but has more difficulties with reproducing the PM10 concentration changes. This is

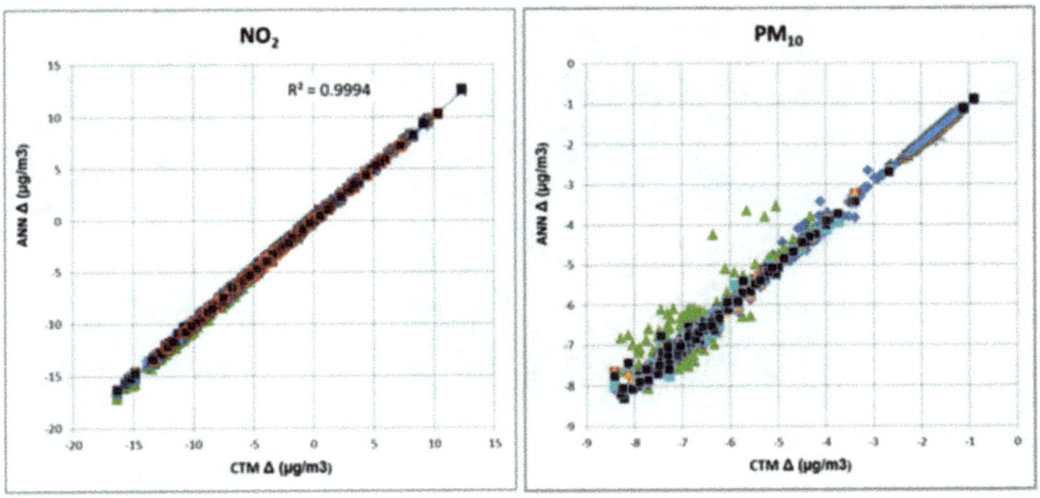

Fig. 5.7 Comparison of ANNs and CTM (AURORA model) results for NO_2 and PM10

especially true for a model run in which only the NOx emissions are changed (represented by small triangles in the figure). For this last case, the average normalised bias amounts to 3.6 % with extreme values of up to 33 % whereas for all the other scenarios the average normalised bias is less than 0.25 %.

5.3.4 *Results*

The application of RIAT+ allows to quickly compute the impacts of any combination of measures. In particular, for BCR, the concentration patterns were examined, together with the distribution of YOLLs assuming a uniform population density in the area.

Even assuming all the measures are in place, the emission changes are limited, and thus, unsurprisingly, also the concentration changes are limited (see Fig. 5.8). The average estimated changes for PM10 are for most measures less than 0.1 μg/m^3. This is within the range of model uncertainty, so these results should be considered with caution. As YOLLs are mainly dependent on PM10 concentration, their value should also be interpreted conservatively. All this said, RIAT+ estimates the impact of all measures to be worth well more than ten million euros per year.

Looking at individual measures, the 'toll' seems the most effective, while other traffic measures, such as LEZ, seem less relevant because a large part of old EURO vehicles will already be replaced by newer types in 2020.

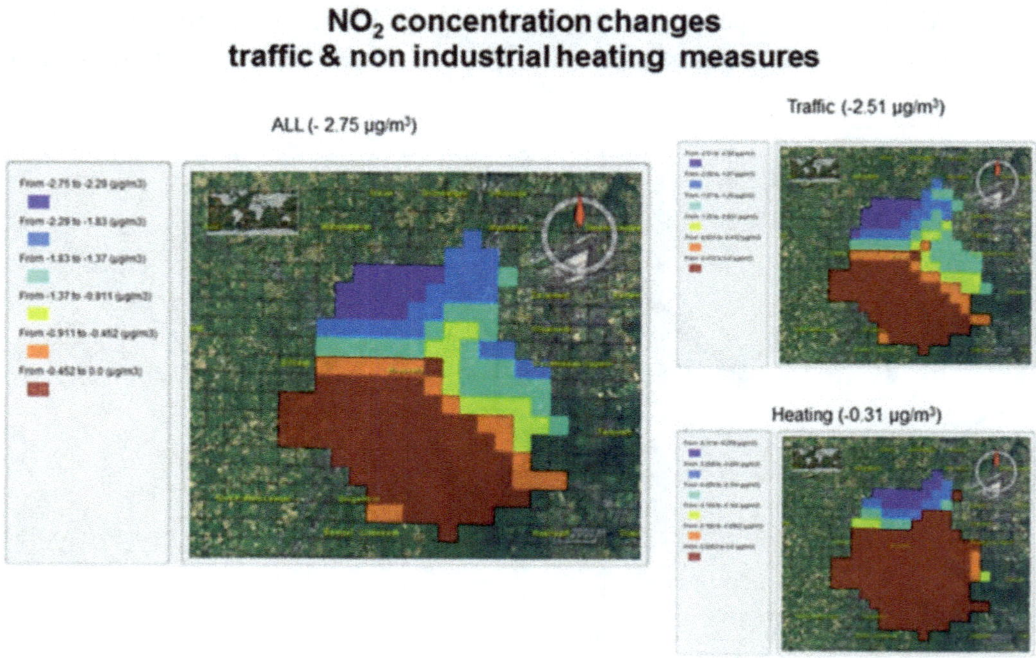

Fig. 5.8 Average NO_2 concentration changes due to the application of different set of measures

5.4 Great Porto Area

The Great Porto Area is a Portuguese NUTS3 (Nomenclature of Territorial Units for Statistics) sub region involving 11 municipalities. It covers a total area of 1024 km^2 with a total population of more than 1.2 million inhabitants. Figure 5.9 shows the location of the Greater Porto Area in Portugal and in its northern region.

This region of Portugal is one of the several EU zones that had to develop and implement an air quality plan (AQP) to reduce PM10. AQP was initially designed using a scenario approach considering the implementation of an a priori defined set of abatement measures (Borrego et al. 2011, 2012). This allowed to identify the most relevant emission sectors: industrial combustion, residential combustion and road traffic.

5.4.1 *Proposed Abatement Measures*

A list of possible abatement measures, including costs and emissions effects was compiled using the GAINS database for Portugal. This database includes: activity details (unabated emission factor, activity level…) and technology details (removal efficiency, CLE and potential application rate, unit cost) for the years 2010, 2015, 2020 and 2025. The reference scenario for March 2013 was considered and 103 «triplets» (sector-activity-technology) were associated to the emission inventory. They are related to: Combustion in energy and transformation industries (SNAP 1) (20 measures); Non-industrial combustion (SNAP 2) (4 measures); Combustion in manufacturing industry (SNAP 3) (23 measures); Production processes (SNAP 4) (7 measures); Solvent and other product use (SNAP 6) (10 measures); Road Transport (SNAP 7) (25 measures); Other mobile sources and machinery (SNAP 8) (14 measures). They are basically all end-of-pipe measures. Technologies for food and drink industry production processes and for construction activities were not included as they are not present in the GAINS database.

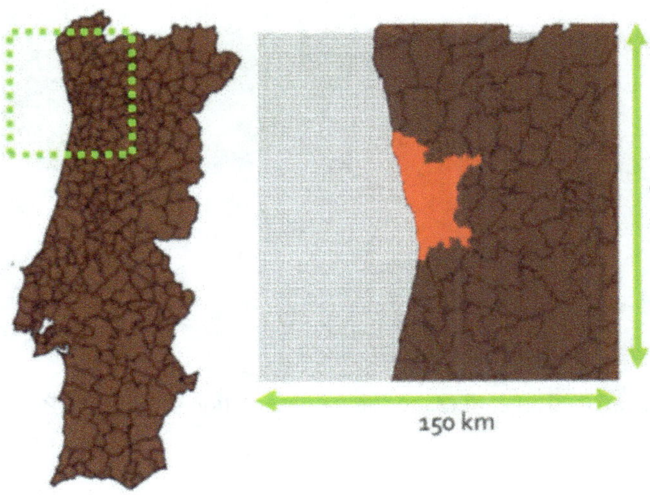

Fig. 5.9 Location of the Great Porto area in Portugal and model grid used

5.4.2 The Chemical Transport Model

The Air Pollution Model (TAPM) (Hurley et al. 2005) was used for the simulation of different mitigation scenarios. It is a 3-D Eulerian model with nesting capabilities, which predicts meteorology and air pollution concentrations. It simulates the transport, dispersion and chemistry of atmospheric pollutants, at both local and regional scale, and it is suitable for long term simulations (e.g. a full year) since it is not strongly time-demanding in terms of computational efforts. Point, line and area/volume source emissions are considered. The model has two components: the meteorological prognostic and the air pollution concentrations component. The meteorological module of TAPM is an incompressible, optionally non-hydrostatic, primitive equation model with terrain-following coordinates for 3D simulations. The results from the meteorological module are one of the inputs to the air pollution component. The gas-phase chemistry mode of TAPM was used, which is based on the semi-empirical mechanism called Generic Reaction Set (GRS), including also the reactions of SO_2 and PM, having 10 reactions for 13 species. The TAPM model was applied to the Great Porto Area (150 km × 150 km) for one entire reference year (2012) with a 2 km by 2 km spatial resolution (see Fig. 5.9) using disaggregated emissions from the Portuguese 2009 emission inventory, which is the most recent available. Its results were compared to the measured values at the monitoring stations inside the model domain. As in Brussels case, we used the methodology proposed by FAIRMODE for the validation.

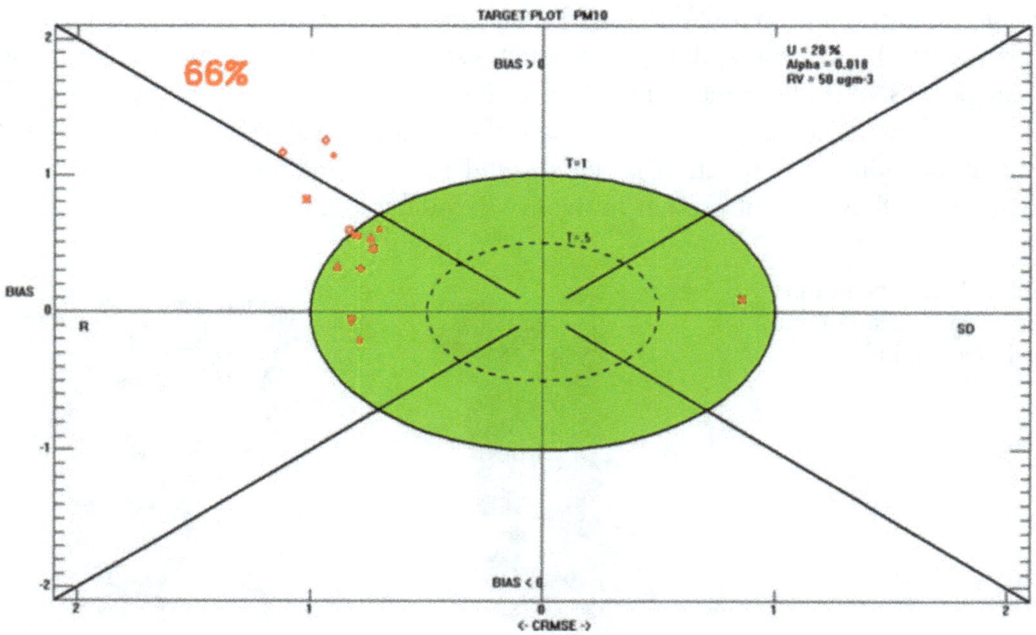

Fig. 5.10 Target diagram for the observation stations for PM10 inside the model domain for 2012

In Fig. 5.10 the target diagram for PM10 results is shown. In this case, modelling results comply at 66 % with the unit circle criterion, even if the overall BIAS is around 33 % of the average value and the average correlation between modelled and actual values is about 0.5.

The four non-complying stations have high values of BIAS and Root Mean Square Error (RMSE), which could be related to an overestimation of background values. The target diagram for NO_2 shows that 69 % of the values comply with the unit circle criterion, while values are similar to PM10 for the other performance indicators.

5.4.3 *Design of Experiments and Source/Receptor Models*

Ten emission sets were defined to train the RIAT+ Artificial Neural Networks for the Great Porto Area. These scenarios have to contain all possible relationships between precursor emissions and the various air quality indices. Ideally, the number of scenarios is determined by checking the incremental improvements to the ANN results of adding additional scenarios to the training dataset.

Starting from the 2009 Portuguese emission inventory, three different emission levels were considered: B (base case), L (low emission reductions) and H (high emission reductions).

The B (base) case considers the evolution of 2009 emissions taking into account the fulfilment of CLE2020 increased by 15 %. The H (high reduction) case is associated to the Maximum Feasible Reduction of emissions in 2020 (MFR2020), further decreased by 15 %. These bounds guarantee that the optimal plausible reductions will lie within those present in the training dataset. The L (low reduction) scenario results from averaging B and H emission values.

The procedure to implement these S/R models requires two steps. In the first step the best ANN structures were chosen on the basis of maximum correlation and minimum RMSE, considering a series of different possible configurations (i.e. different network structure, activation function and number of cells). Then, in a second step the best structure was applied to the whole domain. The quality index considered in this application was the PM10 annual mean. Table 5.3 presents the best ANNs parameters selected for PM10 Neural Network.

To validate the results from the ANN, output values are compared to the results calculated by the CTM. The scatter plot in Fig. 5.11 shows the comparison for an independent validation set which consists of 20 % of the available grid cells not used in ANN training. The good performance of the ANN, with a Normalised RMSE of 0.34 and a correlation coefficient of 0.95 confirms that the ANNs have a sufficient capability to simulate the nonlinear S/R relationship between PM10 mean concentration and the emission of its precursors.

Further analyses confirmed this conclusion. For instance, the average correlation between TAPN and the ANN surrogate model in terms of AQI variations with respect to the base scenario is about 0.93.

Table 5.3 ANN best parameters for PM10 annual mean index

ANNs features	Value
Nodes in the input layer	16
Hidden layer transfer function	Log-Sigmoid
Nodes of the hidden layer	20
Output layer transfer function	Linear
Nodes in the output layer	1
Training function	Levenberg-Marquardt backpropagation
Radius of influence (n° of cells)	4
Training set (n° of cells)	6784
Validation set (n° of cells)	1696

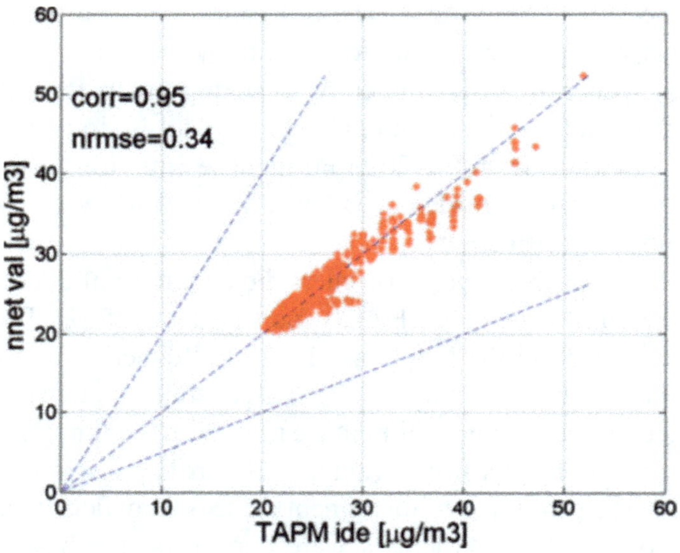

Fig. 5.11 ANNs performances evaluated in terms of scatter plot between ANNs and TAPM results for PM10

5.4.4 *Results*

RIAT+ was applied in the optimization mode and Fig. 5.12 shows the Pareto optimal (efficient) solutions over the Great Porto domain. The horizontal axis of the figure shows the implementation costs (over CLE) of abatement measures expressed in M€, and the vertical axis reports the corresponding efficient AQI value. It shows that a PM10 mean concentration of 28.8 µg/m3 can be reached by adopting emission reduction technologies costing around 7.6 Million € per year (see point C). Points A and Z represent the extreme cases where no actions or maximum feasible reductions are implemented. The other points of the Pareto Curve are intermediate solutions.

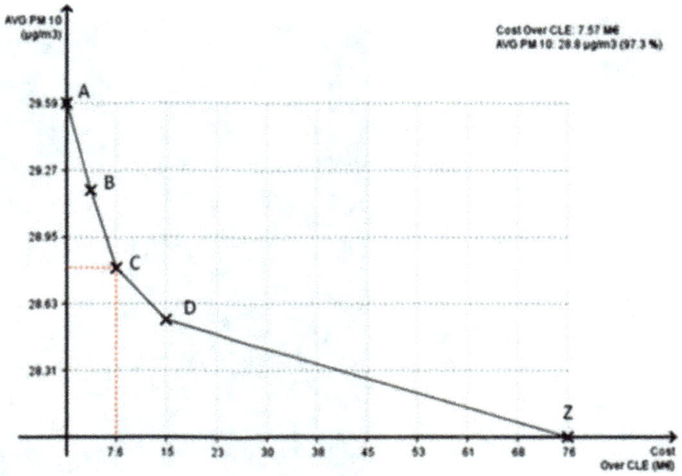

Fig. 5.12 Pareto curve of mean yearly PM10 concentrations

The solution corresponding to point C of the curve, for instance, would be reached mainly acting on non-industrial sector activities (SNAP 2). Road transport (SNAP 7) and other mobile sources and machinery (SNAP 8) could also contribute to the required reduction of PM concentrations. More precisely, the major investment should be in measures related to new and improved fireplaces. These results are consistent with the ones obtained by Borrego et al. (2012): in Portugal, 18 % of PM10 emissions are due to residential wood combustion, which may deeply impact the PM10 levels in the atmosphere. According to the Portuguese emission inventory, this macro sector is the second most important in terms of PM10 emissions, after macro sector 4 (industrial processes), in the Great Porto area.

Figure 5.13 presents the spatial distribution of the expected reductions of PM10 concentration levels, for the Point C of the Pareto curve. The largest reductions of PM10 emissions and concentration levels are expected over the Porto municipality where the population density is higher.

The analysis of RIAT+ results for the selected solution, which implies annual costs around 7.6 M€, shows that some areas can still be expected to exceed the PM10 annual limit value (40 µg/m^3).

Finally, Fig. 5.14 presents the relation between investment cost and benefit measured as reduction of external cost (in term of reduced YOLLs). The ratio between benefits and internal costs significantly decreases when Point B is reached. In other words, the additional gain in health benefits is smaller per additional € invested. However, as it can be seen from the figure, investment costs are always lower than external costs (i.e. below the Y = X line) until point Z. This indicates that acting on emission to reduce PM10 concentrations is always beneficial from a socio-economic point of view.

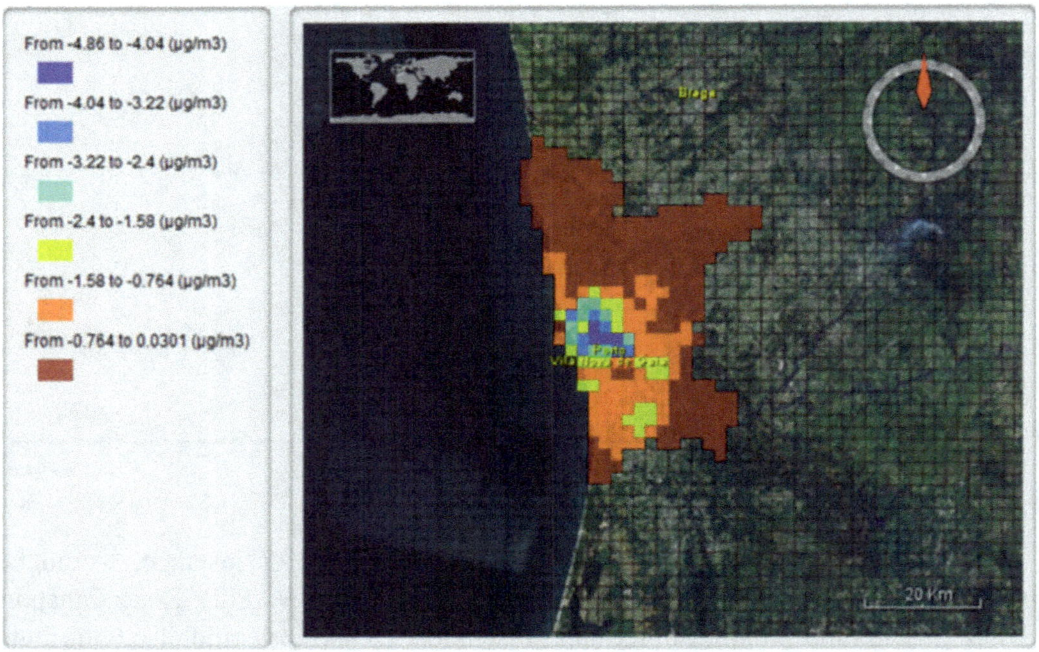

Fig. 5.13 RIAT+ estimated concentration reductions (µg/m^3) correspondent to point C of the Pareto curve

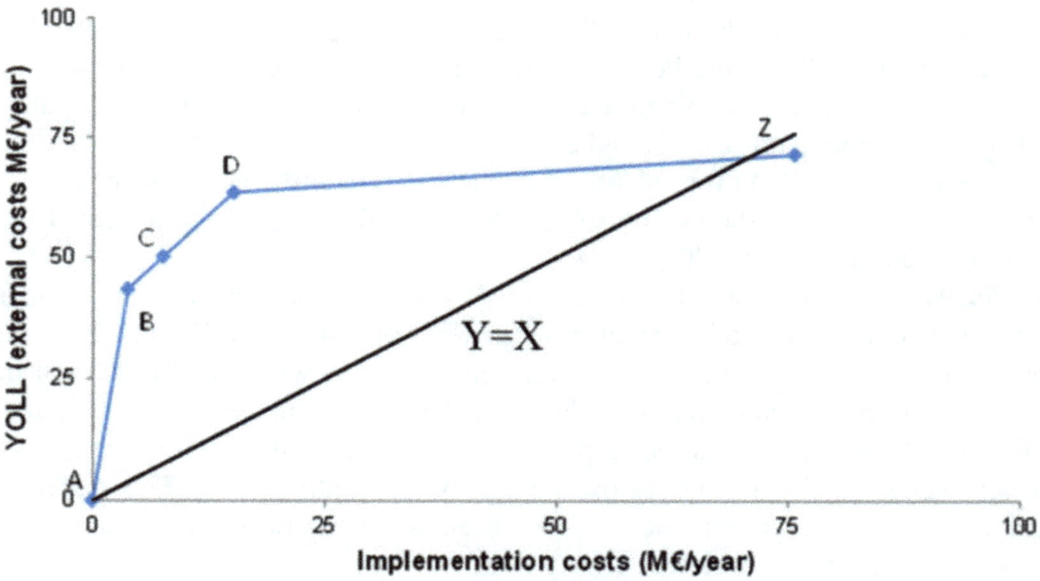

Fig. 5.14 Cost-benefit analysis (implementation vs. external costs)

5.5 Conclusions

From the experiences of application of a comprehensive IAM system (RIAT+) to the test cases of Brussels Capital Region and the Great Porto area a number of conclusions can be drawn.

The list of options for abatement measures is restricted not only by what is technically and economically feasible but possibly even more by political and social acceptance. IAM tools should therefore be further extended to take into account the implications of political and social acceptance at an early stage of the decision process (see also Laniak et al. 2013).

Existing tools can be practically applied in an integrated assessment of air quality not only to consider compliance to the concentration limits but also to efficiently take into account internal and external costs (e.g. health impact) of different available abatement options.

The biggest task when implementing such a comprehensive IAM is—as is also the case in regular air quality modelling applications—to obtain high quality input data on local emissions and the cost and effectiveness of possible abatement measures. When such data is lacking, one can still rely on existing European inventories and databases with data on abatement measures such as EMEP and GAINS well keeping in mind the assumed validity of such data for the region of interest and the implications for the results obtained using the IAM.

If an IAM system uses S/R relationships (artificial neural networks, linear regression, …) to relate emission changes to air quality changes, such relationships should be carefully tested to ensure that they not only correctly replicate the concentration values obtained through more complex modelling tools (e.g. CTMs) but also capture the dynamics i.e. the concentration changes calculated by the model for which they are a surrogate.

In the Brussels case, a lot of effort was put into defining and evaluating specific measures while the impact on air quality of these measures is rather limited due to the dimension of the area selected. A first screening step such as a simple scenario to check the importance of the impacts should be done before using a complex methodology, as the latter has limited added value in such cases.

In the Porto case, a list of available technologies from an existing database was used and the main sectors were selected and identified. Nevertheless, a more local list of measures needs to be decided and discussed with stakeholders and policy makers. With the optimization approach, it was possible to quickly identify the sectors and the entity of optimal investment costs to achieve a given air quality objective and the corresponding benefits.

Acknowledgments This chapter is partly taken from APPRAISAL Deliverable D4.3 (downloadable from the project website http://www.appraisal-fp7.eu/site/documentation/deliverables.html).

References

AIRPARIF (2012) Révision du plan de protection de l'atmosphère d'Île de France (in French: Updating of the air quality plan for Ile de France). http://www.airparif.asso.fr/_pdf/publications/ppa-rapport-121119.pdf. Last accessed Feb 2016

BIM/IBGE (2012) Zich beter verplaatsen in Brussel. 100 tips om het milieu te sparen bij uw verplaatsingen (in Dutch: Improving the way you move around in Brussels: 100 tips to save the environment during you transport). http://documentatie.leefmilieubrussel.be/documents/100conseils_mobilite_NL_LR.PDF. Last accessed Feb 2016

Borrego C, Carvalho A, Sá E, Sousa S, Coelho D, Lopes M, Monteiro A, Miranda AI (2011) Air quality plans for the northern region of Portugal: improving particulate matter and coping with legislation. In: Advanced Air Pollution, Chapter 9, InTech Open Access: F. Nejadkoorki, pp 1–22

Borrego C, Sá E, Carvalho A, Sousa J, Miranda AI (2012) Plans and Programmes to improve air quality over Portugal: a numerical modelling approach. Int J Environ Pollut 48:60–68

Carnevale C, Finzi G, Pisoni E, Volta M, Guariso G, Gianfreda R, Maffeis G, Thunis P, White L, Triacchini G (2012) An integrated assessment tool to define effective air quality policies at regional scale. Environ Model Softw 38:306–315

Carnevale C, Finzi G, Pederzoli A, Turrini E, Volta M, Guariso G, Gianfreda R, Maffeis G, Pisoni E, Thunis P, Markl-Hummel L, Blond N, Clappier A, Dujardin V, Weber C, Perron G (2014) Exploring trade-offs between air pollutants through an Integrated Assessment Model. Sci Total Environ 481:7–16

Cycling England (2007) Valuing the benefits of cycling. A report to Cycling England, May 2007, SQW Group, Cambridge, UK

ECODRIVEN (2008) Campaign catalogue for European ecodriving & traffic safety campaigns. Report from the ECODRIVEN project funded by Intelligent Energy Europe (IEE). http://www.fiaregion1.com/down-load/projects/ecodriven/ecodriven_d16_campaign_catalogue_march_2009.pdf. Last accessed Feb 2016

FEBIAC (2013) Datadigest 2013 'Evolution des immatriculations de voitures neuves par carburant' (in French: Evolution of the registration of new vehicles by fuel type). http://febiac.be. Last accessed Feb 2016

Hurley PJ, Physick WL, Luhar AK (2005) TAPM: a practical approach to prognostic meteorological and air pollution modelling. Environ Model Softw 20:737–752

Laniak GF, Olchin G, Goodall J, Voinov A, Hill M, Glynn P, Whelan G, Geller G, Quinn N, Blind M, Peckham S, Reaney S, Gaber N, Kennedy R, Hughes A (2013) Integrated environmental modeling: a vision and roadmap for the future. Environ Model Softw 39:3–23

Lauwaet D, Viaene P, Brisson E, van Noije T, Strunk A, Van Looy S, Maiheu B, Veldeman N, Blyth L, De Ridder K, Janssen S (2013) Impact of nesting resolution jump on dynamical downscaling ozone concentrations over Belgium. Atmos Environ 67:46–52

Maes J, Vliegen J, Van de Vel K, Janssen S, Deutsch F, De Ridder K, Mensink C (2009) Spatial surrogates for the disaggregation of CORINAIR emission inventories. Atmos Environ 43:1246–1254

Mensink C, De Ridder K, Lewyckyj N, Delobbe L, Janssen L, Van Haver P (2001) Computational aspects of air quality modelling in urban regions using an optimal resolution approach (AURORA). Large-scale scientific computing—Lecture notes in computer science, vol 2179, pp 299-308

Pernigotti D, Thunis P, Belis C, Gerboles M (2013) Model quality objectives based on measurement uncertainty. Part II: PM10 and NO_2. Atmos Environ 79:869–878

STRATEC (2014) Etude relative à l'introduction d'une tarification à l'usage en Région de Bruxelles-Capitale (in French: Study on introducing a toll in the Brussels-Capital region). http://www.stratec.be/sites/default/files/files/C789.pdf. Last accessed Feb 2016

Thunis P, Pernigotti D. Gerboles M (2013) Model quality objectives based on measurement uncertainty. Part I: Ozone. Atmos Environ 79:861–868

Thunis P, Georgieva E, Pederzoli A (2012) A tool to evaluate air quality model performances in regulatory applications. Environ Model Softw 38:220–230

TM-Leuven (2011) Studie betreffende de relevantie van het invoeren van lage emissiezones in het Brussels Hoofdstedelijk Gewest en van hun milieu-, socio-economische en mobiliteitsimpact (in Dutch: Study on the relevance of introducing low emission zones in de Brussels-Capital Region and their environmental, socio-economical and mobility impact). http://www.tmleuven.be/project/lezbrussel/LEZ_BHG_finaal_eindrapport_2011-12-13.pdf. Last accessed Feb 2016

VITO (2011) Potentiële emissiereducties van de verwarmingssector tegen 2030 (in Dutch: Potential emission reductions for the heating sector by 2030). Studie uitgevoerd in opdracht van FOD Volksgezondheid, Veiligheid van de voedselketel en Leefmilieu Vlaamse Instelling voor Technologisch Onderzoek (VITO), Bruxelles

Yarwood G, Rao S, Yocke M, Whitten GZ (2005) Updates of the carbon bond chemical mechanism: CB05, RT-04-00675 prepared for US EPA. Yocke and Co, Novato

Chapter 6
Conclusions: A Way Forward

G. Guariso and M. Volta

Despite a general improvement expected for the next decade in EU, some urban areas and some regions will still struggle with severe air quality problems and related health effects. These areas are often characterized by specific environmental and anthropogenic factors and will require ad hoc additional local actions to complement medium and long-term national and EU-wide strategies to reach EU air quality objectives. These urban areas are also among the territories where most energy is consumed and most greenhouse gases (GHGs) are emitted.

So far, abatement strategies for air pollution and GHGs have mostly been treated separately. However, increasing scientific evidence shows that air pollution and climate change policies must be integrated to achieve sustainable development and a low carbon (LC) society. Combined efforts to deal with air pollution and climate issues at the urban level will be particularly important because here is where most people are exposed to air pollution, and 75 % of global GHG emissions are generated (Schmale et al. 2014; UNEP and WMO 2011). Urban air pollution and climate change simultaneously posing a serious threat to citizen longevity and quality of life: reducing and mitigating the corresponding impacts is vital. Such integration may exploit known synergies and would lead to substantial cost savings and important benefits for human health and the environment.

A widespread application of classical end-of-pipe measures, such as various types of filters and catalysts, may provide only a moderate improvement to the air quality in cities and regions and will not reduce their GHG emissions. Until now, however, the modelling and analysis underpinning the development of abatement strategies in the EU and in the Convention on Long-Range Transboundary Air Pollution (CLRTAP) has focussed on technical measures. In the future, non-technical mea-

G. Guariso
Politecnico di Milano, Milan, Italy

M. Volta (✉)
Università degli Studi di Brescia, Brescia, Italy
e-mail: marialuisa.volta@unibs.it

G. Guariso and M. Volta (eds.), *Air Quality Integrated Assessment*,
PoliMI SpringerBriefs, DOI 10.1007/978-3-319-33349-6_6

sures will become increasingly important for the mitigation of both air quality and climate change and policy development will require analytical tools which are capable of dealing with a wider range of measures and changes, which improve energy efficiency and air quality at the same time, while at least maintaining the same level of services. These measures cover a very wide spectrum, going from the change of production processes in industry, to shifts in transport modes, to changes in buildings, urban structures and plans, and in citizens' way of life.

Such important structural societal changes will require the involvement of a larger range of actors, with the citizen as a key stakeholder (Laniak et al. 2013). In many cases, citizens will need to modify their perceptions and behaviour: a process that may take much longer than the adoption of a new technology or a new regulation. Even when all the impacts are accounted for in a scenario analysis or in an optimization procedure, the final output of current tools is a classical "plan", i.e. a set of measures to be implemented, by a possibly "almighty" decision-maker. These plans disregard crucial issues, notably how these measures might be accepted by citizens, how long it will take to implement them and through what normative, economic, or simply persuasive ways they can be actuated. This, on the contrary, requires a long-term perspective and it is not sufficient to state a traditional air quality and GHG emission reduction plan with fixed actions and fixed targets to be achieved in a given number of years. What is needed is a continuous process, which takes into account the continuous changes of society and evolution of technology, as well as external conditions (such as, for instance, international agreements on climate or on transboundary pollution) and tries to accompany and foster the transition of society toward cleaner air and lower carbon emissions.

City and regional authorities cannot impose such a transition, particularly if behavioural changes are required, but they can certainly influence its direction and speed. Managing this transition is a complex task for city and regional authorities, that requires taking into account the governance levels to correctly target the areas where control measures should be implemented for highest efficiency and, at the same time, requires being able to confront environmental equity issues, public criticism and even protest.

Increasing awareness of this complexity has triggered the development of the "Transition Management (TM)" concept. Pioneered at the end of the '90s in the Netherlands in sectors such as water management, energy supply or mobility (e.g. Rotmans et al. 2001), it has been taken up in the meantime in several other countries and by international organisations such as the OECD (2014) or EEA (2014) in order to guide and inform innovation policy and associated sectoral policies.

Transition Management has also been proposed as guiding strategy for tackling major societal challenges in other areas (CEC 2011). Even if the European Commission suggests an evolving approach for moving to a competitive low carbon economy (EC 2011, EU 2013), in the development of plans for cleaner air together with low GHG emission, TM can be considered as quite new approach.

To apply such a new concept, decision makers must be supported by a systematic framework that can be adapted to the specific circumstances of the different regions and cities and to the complex systems dynamics of societal and technological changes, as well as by a working set of tools that may help at different stages of the transition.

Indeed, TM includes and extends the concept Integrated Assessment Modelling, a methodology that, as seen in the preceding chapters, has been developed over the last two decades to select "optimised" policies aiming at reducing the negative impacts of air pollution and climate change.

Central to the concept of TM is a multi-level perspective on long-term change processes (Rotmans et al. 2001) that may take a long time, sometimes decades, to be realised. Although there may be periods of slower and faster development, in general, there are no major jumps due to the manifold inter-dependencies in socio-technical systems. TM is therefore a goal-oriented process of continuous learning and adjustment among a broad range of actors and stakeholders.

The application of TM to air quality and at climate change mitigation actions will need the design of innovative strategies based on an in-depth analysis of the scientific findings (e.g. atmospheric composition dynamics, new modelling approaches, technological innovation, social analysis) and on technical and economic challenges (e.g. implementation of cleaner technologies or urban planning). It will also need the definition of the right level of actions to find efficient synergies and good compromises between European/national/local policies and air quality and climate issues.

While consensus can often be reached on the overarching transition goals (e.g. cleaner air in cities, fighting climate change), conflicts of interest may easily arise once those objectives and targets become more specific, and when policies in one area have negative impacts on another. Specific policies thus need to be carefully designed in order to disentangle the mechanisms behind acceptability for the different actors and stakeholders concerned and to avoid disruptive conflicts. Negotiating and moderating debates about conflict-prone policies is crucial to the success of TM (Smith et al. 2005). In order to underpin these debates and decision-making processes with information and knowledge as accurate as possible, suitable tools and approaches for data collection, analysis and assessment are needed. They are also essential to enable monitoring, learning, and adjustment during the relatively long time periods of the TM processes.

A key issue in the development of a transition process is the assessment of the social acceptability of political decisions. Different techniques are already available for this purpose.

Discrete Choice Models, for instance, present the advantages of stressing the trade-offs among different choice alternatives and have been used for the first time within the SEFIRA FP7 coordination action (http://www.sefira-project.eu) to assess the acceptability of Air Quality and Climate Policies. Different degrees of acceptability depend on citizens' preferences and their awareness of the drivers/pressures/impact in AQ and LC. These perspectives may be investigated using a discrete choice analysis performed asking citizens to fill a traditional questionnaire on their choices in relation to AQ and LC policies and/or developing CAWIs (Computer Assisted Web Interviewing). When conducted in selected regions where specific actions have already been applied, these methods allow to better investigate the impacts of local and regional policies on regional socio-economical systems and on how such impact is accepted by individuals. Two scenarios will be possible: (i) different local/regional environmental policies/measures having the same impacts can be implemented according to the individual preferences (acceptability ranking); (ii) most effective policies do not respect the acceptability ranking or, in the worst scenario are not accepted at all: in this case, communication plays a key role in building awareness on the trade-off existing between people desires and environmental constraints. This requires the development of education (long term perspective) and communication (short term perspective) tools, with a focus on tools aimed at raising awareness on win–win strategies for AQ and LC. Also the presence and impacts of AQ and LC policies in social media may contribute to the acquisition of scientifically ground knowledge that can be translated into concrete everyday life choices by citizens.

The definition of TM goals and vision requires a very strong involvement of all stakeholders, and needs to be embedded in a political process making commitment. An iterative process of monitoring, learning and adjustment must be designed, involving these stakeholders at regular intervals.

The experiences gained and the lessons learned must be suitably codified for subsequent use by other cities and regions; for instance, by compiling and circulating a detailed and comprehensive guidebook. This approach is similar to the provisions of the SEA (Strategic Environmental Assessment) Directive 2001/42/EC.

Research should be aimed at defining how to co-design and set up a suitable "toolbox" to support decision makers in selecting the measures and strategies that address their transition goals to be implemented. Such a toolbox should include models, databases, guidelines, dissemination formats, and the supporting software. The tools could be identified and structured on the basis of the DPSIR methodology that, as shown in the previous chapters, is well suited for describing the interactions between society and the environment.

Monitoring tools are also needed to continuously assess the effectiveness of the AQ policy in reaching its goals, in particular the socio-economic consequences.

All the tools should be connected within a suitable ICT environment.

Two components of the ICT infrastructure deserve a special attention, given that research activities in the recent past have mainly focused on the development and application of European and regional models: the first is the creation and diffusion of a (meta)data dictionary; the second, the implementation of a common database for emission abatement measures.

The dictionary can be implemented in different ways, including an ontology, to ease the communication between all the involved parties: scientists, decision-makers, citizens, all other stakeholders and within each group. Its purpose is to clearly define all the variables used in the DPSIR scheme in such a way that their meaning can be precisely understood by all as well as other perspective stakeholders both from a qualitative and quantitative viewpoint. It is in fact common that, for instance, emissions are classified and/or measured in different ways in different sites;

or actions, named in the same way, are in practice actuated with different means. The dictionary will clarify these differences thus allowing consistent comparisons between different situations and clear definitions of best practices.

The database of emission abatement measures should contain a standardized description of different reduction activities, as it is currently the case of end-of-pipe measures in the GAINS database. This means that all the activities must be accurately described with some indication of their cost, evolution, and effectiveness. This again would allow an easier identification of best practices as well as of the (private or social) investment needed for their implementation.

Finance and economic policy measures (public–private partnerships, concessional grants and loans at city, regional, national and EU levels, full private sector financing), for instance, can be used to facilitate rapid market deployment for innovative abatement solutions. In this respect, case studies of effective economic policy instruments to reduce air pollution and carbon emissions in European cities in the transport and heating sectors are already available (see, for instance, Chap. 5) as well as experiences in the adoption of economic policy instruments such as regulation, pricing, public funding, subsidies and exemptions.

Other measures can be related to educational efforts as integrated parts of air pollution and low carbon policies. These measures should address the challenges of consumer-citizens involvement and learning.

Once again, sharing studies, costs, results, and acceptance of these actions within a common framework over the Internet would allow for intercomparison and mutual learning among all local governments and environmental agencies.

References

CEC (2011) COUNCIL OF THE EUROPEAN UNION, The Budapest Declaration, Transition towards sustainable food consumption and production in a resource-constrained world. 10138/11

EC (2011) A Roadmap for Moving to a Competitive Low Carbon Economy in 2050, COM/2011/112 final

EEA (2014) Multiannual Work Programme 2014–2018: Expanding the knowledge base for policy implementation and long-term transitions, Luxembourg

EU (2013) Regulation (EU) No 1300/2013 of the European Parliament and of the Council of 17 Dec 2013

Laniak GF, Olchin G, Goodall J, Voinov A, Hille M et al (2013) Integrated environmental modeling: a vision and roadmap for the future. Environ Model Softw 39:3–23

OECD (2014) Synthesis report on system innovation, Paris. https://www.innovationpolicyplatform.org/sites/default/files/general/SYSTEMINNOVATION_FINALREPORT.pdf. Last accessed Mar 2016

Rotmans J, Kemp R, van Asselt M (2001) More evolution than revolution: transition management in public policy. Foresight 3:15–31

Schmale J, Shindell D, von Schneidemesser E, Chabay I, Lawrence M (2014) Air pollution: clean up our skies. Nature 515:335–337

Smith A, Stirling A, Berkhout F (2005) The governance of sustainable socio-technical transitions. Res Policy 34:1491–1510

UNEP, WMO (2011) Integrated assessment of black carbon and tropospheric ozone, Nairobi. http://www.unep.org/dewa/Portals/67/pdf/BlackCarbon_report.pdf. Last accessed Mar 2016

www.ingramcontent.com/pod-product-compliance
Lightning Source LLC
LaVergne TN
LVHW082239200926
848484LV00008B/2376
* 9 7 8 1 0 1 3 2 6 7 7 7 2 *